| 빛의 디자인 | DESIGN OF LIGHTS |
| 공간의 마술 | SPATIAL MAGIC |

빛: **THE LIGHT:**   이연소   **LEE**
기본으로 돌아가다   a return to the essence   지음   **YEON SO**

夜 밤은 나에게
생명의 빛을 일으키게 했다.

사람들이 빛을 환상이라고 말하지만,
나에게 빛은 내면에서 투사된 우주이다.

판테온 | Pantheon

NIGHT

The night gave me
the light of life.

People say light is an illusion,
for me, light is the universe projected from within.

誠

밤의 어둠 속에서 한 줄기 빛이 공간의 가치를 만들어 낸다. 빛 연출은 촛불 하나 밝히듯 작고 조용하다.

국가유산은 우리의 정신문화이고, 빛이 그 가치를 드러낸다.

서울 창덕궁 달빛기행에서 빛 연출은 궁궐 본연의 가치와 품격을 보여주는 것이다.

창덕궁에 보름달이 뜨고, 실내에 설치된 빛이 창살로 새어 나와 달빛에 젖어 고요하다.

빛의 디자인, 세계문화유산 창덕궁 달빛기행
Light Design of Moonlight Tour of Changdeokgung Palace, 2013

SINCERITY

In the darkness of night, a beam of light creates the value of a space. Like lighting a single candle, the light production is small and quiet. National heritage is our spiritual culture, and light reveals its value. In the moonlight tour of Changdeokgung Palace in Seoul, the light production shows the value and dignity of the palace itself. A full moon rises over Changdeokgung Palace, and the light from the inside shines through the grate, creating a moonlit and serene atmosphere.

花 거제시 장승포항에 해가 지면,
빛의 동백꽃이 곱게 피어난다.
장승포를 찾은 사람들 가슴에도 동백꽃이 피어난다.

빛의 디자인, 거제시 장승포항 | Light Design of Jangseungpo Port, Geoje City, 2021

FLOWER
As the sun sets on Jangseungpo Port in Geoje City,
the light camellia blooms beautifully.
Camellias bloom in the hearts of those who visit Jangseungpo Port.

風

의암호수를 건너온 바람이 의암댐을 푸른 빛으로 그려낸다.
호반의 도시 춘천 의암댐에 밤이 오면
빛 연출로 물과 바람이 푸른 빛으로 피어난다.

빛의 디자인, 춘천시 의암댐 | Light Design of Uiam Dam, Chuncheon City, 2021

WIND
The wind from across Uiam Lake paints the Uiam Dam in blue.
When night falls at Uiam Dam in the lakeside city of Chuncheon,
through light production, the water and wind bloom with blue light.

머리글

빛과의 동행

바로크시대(Baroque Art) 빛의 화가인 렘브란트(Rembrandt Harmenszoon van Rijn, 1606~1669)와 인상주의 화가인 르누아르(Pierre Auguste Renoir, 1841~1919)의 그림 앞에 서면 내 안에 웅크리고 있던 뭔가가 꿈틀대며 살아난다. 캔버스에 빛 방울이 쏟아져나와 나를 휘어잡는다. 나는 나에게 영감을 주는 화가들의 그림은 물론, 클래식 음악, 연극, 영화와 뮤지컬을 두루 좋아했다. 젊은 날에 대학로와 인사동의 다양한 예술 장르를 접한 것은 뒷날 빛 연출 총감독의 바탕이 됐다.

내 연구실 벽에 걸린 오병욱 작가의 「내 마음의 바다(Sea of My Mind)」가 나를 내려다보고 있다. 이 그림은 물감을 캔버스에 직접 뿌려가면서 작업했는데, 물감 하나하나 자잘한 빛 알갱이가 어우러져 푸른 바다가 되어 나를 환상의 세계로 데려간다. 때로는 바다가 신비한 우주 공간이 되기도 했다. 내가 꿈꾸는 빛의 세계도 이렇게 작은 빛 방울이 모여 거대한 빛의 바다 혹은 빛의 산맥이었다. 나에게 빛은 늘 신비한 우주였다.

PREFACE

Walking with the light

I receive a strong reaction when I stand in front of the Baroque Art paintings of Rembrandt Harmenszoon van Rijn (1606-1669) and Impressionist painter Pierre Auguste Renoir (1841-1919). The light in the paintings capture my attention and stirs something inside me. I am a fan of classical music, drama, movies, musicals, and paintings that inspire me. I was exposed to diverse artistic genres in Daehak-ro and Insadong, which laid the foundation for my later work as a light production director.

I have Byungwook Oh's "Sea of My Mind" painting staring down at me from the wall in my studio. The artwork is created by spraying paint directly onto the canvas, with each drop of paint forming a blue sea of light. It transports me to a fantasy, where sometimes the sea transforms into a mysterious outer space. In my dreams, the world of light is a vast sea of light or a mountain range formed by tiny droplets of light. For me, light has always been a mysterious universe.

나는 빛을 디자인하는 빛 연출가(Light Artist)이자 빛 연출 총감독(Lighting Director)이다. 나의 무대는 온 세상이며, 무대에 자아의 빛을 투사한다. 나는 나의 빛 작품을 대하면 가슴이 벅차오른다. 거기에는 내 영혼이 서려 있기 때문일 것이다. 그러나 나는 빛의 무대에서 내려오면 공허해지고, 내 안에 웅크리고 있던 뭔가가 다시 살아나 나를 빛의 우주로 데려간다. 나에게 빛은 동반자이자 숙명이었다.

나는 여기에 빛의 물감을 이렇게 사용했다는 이야기와 사람들에게 들려주고 싶은 빛의 이야기를 담았다.

지금까지 나를 아끼고 응원해 주신 최기수, 이윤호 두 교수님께 감사드린다. 그리고 항상 나와 함께한 ㈜유엘피 「좋은 빛 디자인연구소」 스태프의 도움이 있었기에 지금의 내가 존재할 수 있다고 생각한다. 내가 걸어가는 길목을 환한 빛으로 지켜주신 아버지와 어머니, 그리고 사랑하는 아내 희우와 딸 수민에게도 고맙다는 말을 전한다.

2025년 3월
이연소

I am a light artist who designs light and a lighting director who directs light in a space. My stage is the world, and I project the light of self onto it. I get excited about my light creations, for therein lies my soul. But when I step down from the stage of light, I feel empty, and something that has been dormant within me comes to life again and takes me into the universe of light. For me, light is a companion and a destiny.

Here's the story of how I used the paint of light and the story of light that I want to tell the people.

I am grateful to Professor Key Soo Choi and Professor Yoonho Lee for their support and encouragement, and to the staff of the ULP Co. Ltd., 「Good Light Design Research & Development Center」 who have always been with me. I would also like to thank my father and mother, my beloved wife, Hee Woo, and my daughter, Soo Min, for their support and guidance.

March 2025
LEE YEON SO

차례

	머리글	010

Chapter I

빛을 보다: 夜

밤 · 夜	020
경 · 景	026
빛 · 光	032

Chapter II

빛을 느끼다: 景

빛 연출 정의   042

빛 연출 대상   044

빛 연출 이미지   048

    부드럽고 따스한 빛   050

    맑고 시원한 빛   051

    생동감을 주는 빛   052

    마음을 담는 빛   053

    살아있는 빛   054

    성스러운 빛   055

빛 연출 원칙   062

    따뜻한 빛, 평온함을 느낀다   064

    부드러운 빛, 아름다움을 밝힌다   066

    그곳만의 빛, 이야기를 담는다   068

빛의 도시   072

    꺼지지 않는 빛의 도시, 홍콩   074

    빛이 절제된 품격의 도시, 파리   076

    따뜻한 마음을 전하는 도시, 리옹   078

    감성의 빛을 담아내는 도시, 시드니   080

Chapter III

빛을
담다: 光

밤을 비우다: 아름다운 한강의 밤 '한강 빛의 르네상스'	088
한강르네상스 빛의 마스터플랜	090
서울시 여의도한강공원 빛 연출	114
서울시 난지한강공원 빛 연출	124
자연을 담다: 호수와 대나무 숲 그리고 라운지 공간	128
경기도 광교신도시 호수공원 빛 연출	130
전라남도 담양 죽녹원 빛 연출	138
인천국제공항 아시아나항공 라운지 빛 연출	146
시간을 넘어서다: 유네스코 세계유산 궁궐과 건축물	162
경복궁, 궁중문화축전의 빛을 밝히다	166
창덕궁 달빛기행, 빛으로 꽃피우다	184
운현궁, 흥선대원군의 이상(理想)을 빛으로 채우다	212
공간을 빛나게 하다: 빛과 소리로 충만한 풍경	224
원주시 소금산 그랜드밸리 '2021 나오라쇼'	226
거제시 장승포항, 동백꽃 붉은빛으로 피다	240
창경궁 대춘당지 '물빛연화'	252
빛을 밝히다: 장소성을 담아내는 영롱한 빛	270
석촌호수, 물의 모습을 빛과 소리로 담다	272
서해에서 불어오는 바람을 빛으로 느끼다	274
제주도 서귀포 생태공원 달빛과 별빛을 품다	276
서울, 빛의 혈(血)을 발견하다	278

Chapter IV

빛 연출과
공간 마술사: 誠

감성을 담은 빛 연출	284
빛 연출의 진정한 가치	286
빛 연출가가 되려는 이에게	292
빛 연출가 이연소가 걸어온 길	294
참고문헌	296
이미지 출처	298

TABLE OF CONTENTS

	Preface	010

Chapter I
SEE THE LIGHT:
NIGHT

	Night	020
	Soul of Nightscape	026
	Light	032

Chapter II
SENSING THE LIGHT:
SCENERY

	Definition of Light Production	042
	Object of Light Production	044
	Images of Light Production	048
	Soft and Warm Light	050
	Clear and Cool Light	051
	Vibrant Light	052
	Light Filled with Emotion	053
	Energetic Light	054
	Sacred Light	055
	Principles of Light Production	062
	Warm Light Feels Peaceful. 「Safety」	064
	Soft Light Illuminates Elegance. 「Beauty」	066
	Unique Light Tells the Story. 「Identity」	068
	City of Light	072
	Hong Kong, the City Lights Never Goes out	074
	Paris, the City of Dignity with Understated Light	076
	Lyon, the City that Gives Warm Heart	078
	Sydney, the City that Captures Light of Emotions	080

Chapter III
CAPTURING THE LIGHT: LIGHT

Empty the Night: Light Master Plan of 'Hangang River Renaissance'	088
Light Master Plan of Hangang River Renaissance	090
Light Production of Yeouido Hangang Park in Seoul, Korea	114
Light Production of Nanji Hangang Park in Seoul, Korea	124
Embrace Nature: Lake, Bamboo Forest and Lounge Area	128
Light Production of Gwanggyo New City Lake Park, Gyeonggi-do	130
Light Production of Juknokwon Garden in Damyang, Jeollanam-do	138
Light Production of Asiana Airlines Lounge at Incheon International Airport	146
Beyond Time: UNESCO World Heritage Palaces and Architecture	162
Gyeongbokgung Palace, Illuminating the Palace Cultural Festival	166
Changdeokgung Palace, Walking in the Enchanting Beauty of Moonlight	184
Unhyeongung Palace, Filling the ideals of Heungseon Daewongun with light	212
Make the Space Shine: Landscapes Filled with Light and Sound	224
'2021 Naora Show', Wonju Sogeumsan Mountain's Grand Valley in Wonju City	226
Camellia Blooms in Red, Jangseungpo Port, Geoje City	240
'Mulbityeonhwa', Daechundanji Pond, Changgyeonggung Palace	252
Let the Light Shine: Brilliant Light that Capture a Sense of Place	270
Seokchon Lake, Capturing the Appearance of Water through Light and Sound	272
Feel the Breeze Blowing from the West Sea through Light	274
Embracing Moonlight and Starlight, Seogwipo Ecological Park, Jejudo Island	276
Seoul, Discovering the Veins of Light	278

Chapter IV
LIGHT CONDUCTOR AND SPATIAL MAGICIAN: GENRAL LIGHT DIRECTOR

The Light Production with Emotions	284
The True Value of Light Production	286
For Individuals who Aspire to Become a Lighting Director	292
The Path of Light Director Lee Yeon So	294
References	296
Photo Credit	298

빛은 도시의 혼과 가치를 담아내는 예술이다.
단순히 어둠을 밝히는 조명이 아닌, 감성을 머금은 차별화된 빛은 공간의 고유한 이야기를 펼쳐낸다.
그 빛이 만들어낸 서사는 도시의 매력을 극대화하며, 사람들의 마음 깊은 곳에 잔잔한 울림을 남긴다.
진정한 빛은 단순한 장식이 아닌 공간과 사람을 따뜻하게 이어주는 감동의 매개체이다.

빛의 마스터플랜, 서울시 빛공해 방지계획
Light Master Plan of Light Pollution Prevention Plan, Seoul, 2015

Light is an art that captures the soul and value of a city.
It is not merely lighting that brightens the darkness, but differentiated light imbued with emotion that unfolds the unique story of a space. The narrative created by that light maximizes the city's charm and leaves a quiet resonance deep in people's hearts.
True light is not just decoration; it is a medium of emotion that warmly connects space and people.

Chapter I
빛을 보다:
夜

SEE THE LIGHT: NIGHT

밤은 밤다울 때 가장 아름답다

밤은 어둠 자체를 보존하고 배려할 때, 밤 속에서 작은 빛이 아름다울 수 있다. 밤을 밝히려고 무절제하게 빛을 과도하게 사용하면 밝은 것도 아니고 어두운 것도 아닌 회색빛이 된다. 빛을 디자인할 때는 밤의 감성을 존중하는 마음이 빛 연출의 시작점이 된다.

The Night is Most Beautiful When It is Dark

A little light can be beautiful in the night, when it is preserved and treated with respect for the darkness itself. When light is used excessively to illuminate the night, it creates a gray light that is neither bright nor dark. Respecting the sensibility of the night is the starting point for designing light.

밤 · 夜

밤은 비움의 시간이다.
나에게 밤의 시간과 공간은 비어 있는 무대와 같다. 연극 공연을 관람할 때, 무대가 모두 밝아졌을 때는 무대의 깊이감이나 그 내용을 정확하게 알 수가 없다. 그러나 공연 시작을 위해 전체 공간이 어두워지고 주연(主演)을 밝혀주는 밝은 빛과 조연(助演)을 비추어 주는 빛, 배경(背景)을 만들어 주는 전반적인 빛의 위계에서 관객은 비로소 극의 상황 속으로 빠져들게 된다. 내가 밤의 어둠을 매력의 공간으로 여기는 이유는 많은 이야기를 풀어낼 무대이기 때문이다.

NIGHT

Night is a Time for Emptiness.
The stillness of night feels like an empty stage for me, where time and space intermingle. When the stage is all lit while you watch a theater performance, you do not really know the depth of the stage or what is going on. However, when the entire space is darkened for the start of the performance, the audience is drawn into the situation of the play by the hierarchy of light: bright lights that illuminate the main characters, lights that illuminate the supporting characters, and overall light that creates the background. The reason I find the darkness of the night so fascinating is that it is a stage which many stories can be told.

밤 속에 빛나는 별빛 | Starlight in the night

빛 연출가는 밤 자체의 속성인 어둠을 소중하게 생각한다.

나는 "밤은 밤다운 어둠이 담겨있어야 한다."라는 말을 자주 한다. 우리가 맑고 깨끗한 밤의 이미지를 보전하고 지켜나갈 때, 밤의 장소, 그 공간만의 이야기를 담백하게 담아낼 수 있기 때문이다.

   밤의 감성을 배려하는 좋은 빛 디자인은 공간의 장소성을 담아내는 빛을 연출할 때이다. '장소'의 사전적 의미로는 어떤 일이 이루어지거나 벌어지는 곳, 지리적 위치나 지리적 공간을 가리키는 말이다. '장소성'이란 장소가 지니는 본질로서, 장소가 지니는 고유한 의미, 사람들의 경험이나 공유를 통해 사람이 갖게 된 인식과 의식을 포괄한다.

공간을 기획하고 설계하는 사람에게 고민을 많이 하게 하는 개념이 장소성이다. 장소마다 독특한 본질과 성격이 있는데, 그것을 어떻게 표현해야 하나, 이 장소에 어떤 시간의 흔적이 숨어 있나, 이를 어떻게 발굴해야 할까 등 다양한 문제와 마주하기 때문이다. 마치 깊이 묻혀있는 역사적인 유물을 탐색해 나가는 고고학자처럼 많은 연구와 고민을 수반하게 되며, 전문성을 바탕으로 한 높은 차원의 예술적 상상력까지 요구되기도 한다.

Light Directors Value Darkness as an Attribute of Night Itself.

I often say, "A night should contain night-like darkness." When we preserve and protect the image of a clear, pristine night, we are able to tell the story of a place at night, a story that is unique to that space.

   The dictionary definition of 'Place' is a space where something happens or takes place, a geographical location or geographical space. 'Placeness' is the essence that encompass the unique meaning of a place, the perception and consciousness that people have of it through experience or sharing.

Place is a concept that causes a lot of thought for those who plan and design space. Each place has a unique essence and character, and we are faced with various issues such as how to express it, what traces of time are hidden in this place, and how to excavate them. Like an archaeologist searching for deeply buried historical artifacts, it involves a lot of research and thought, and often requires a high level of artistic imagination as well as expertise.

요즘 밤은 과거보다 많이 밝아졌다. 인공조명 남용으로 일부는 훼손된 밤의 이미지 같은 부작용도 생겨났다. 예컨대, 서울의 대표적 상권인 동대문 주변의 쇼핑단지는 낮보다 밤이 더 밝다. 동대문 상인이나 쇼핑단지를 찾는 사람들에게는 좋을지 모르지만, 주변에서 생활하는 주민들에게는 불편한 게 사실이다. 쇼핑단지의 어두운 공간을 밝히기 위해 한껏 고개를 하늘로 향한 인공조명은 빛이 하늘을 향하고 있는데, 이 빛이 도시의 구름과 먼지층에 반사되어 밤하늘이 밝아지는 일부 백야현상(White Night)이 나타나기도 한다. 결과적으로 과도한 인공조명이 주변 주민들의 숙면을 방해하는 요소가 된다.

이처럼, 좋은 빛 디자인도 과하면 서로 경쟁하게 되고, 경쟁에 치우치면 밤 본연의 모습이 사라진다. 그동안 동대문 주변의 쇼핑몰(Shopping Mall)은 손님들 눈에 잘 보여야 했기 때문에 서로 경쟁하듯 밝아졌고, 이렇게 빛의 폭탄 속에서 마냥 살아야 하는지 냉정하게 돌아보아야 할 때이다. 시대가 변화하더라도 바뀌지 않아야 할 중요한 것은 밤 자체의 어두운 환경을 먼저 배려하는 것이고, 밤 본연의 이미지를 훼손해서는 안 된다는 뜻이다.

Nights are much brighter these days compared to the past due to the overuse of artificial lighting. This has resulted in some side effects, such as a disrupted nighttime environment. For instance, the shopping complexes around Dongdaemun, Seoul's main commercial center, are now brighter at night than during the day. While this may benefit the merchants and visitors, it causes discomfort for the residents living around the area. The artificial lights are directed upwards to illuminate the dark spaces of the shopping complex, and the light is then reflected off the clouds and dust layer of the city, resulting in excessively bright nights. As a result, the excessive artificial light can disrupt the sleep of nearby residents.

In the same way, good light designs can compete with each other when they are overdone, and when they become too competitive, the essence of the night is lost. Shopping malls around Dongdaemun have been brightened to compete with each other because they need to be visible to customers, and it is time to reflect on whether we should live in such a luminous blast. The important thing that should not change even as the times change is to consider the dark environment of the night itself first, and not to damage the image of the night itself.

한강 두물머리의 따듯한 일출 | Warm sunrise of the Duemulmeori, Hangang River

빛은 마음으로 느끼는 것이다.

빛을 연출하기에 앞서서, 빛에 대해 다음 세 가지를 염두에 둔다. 첫째, 사람들의 안전한 삶에 도움을 주는 빛이다. 밤에 발생할 수 있는 위험, 불안감이나 위기감으로부터 사람을 보호하는 안전한 빛이다. 아름다운 밤을 연출할 때조차도 안전이 우선돼야 한다. 둘째, 밤 본연의 모습을 보여주는 빛 연출이다. 인공적인 조명으로 무조건 밝게 하는 것은 바른 빛 연출이 아니다. 밤은 어둠을 담은 고유의 시간이기 때문에 밤 본연의 모습을 지켜주는 '배려하는 빛'을 염두에 둬야 한다. 셋째, 장소성을 담아내는 절제된 빛의 연출이다. 절제된 빛 연출이란 공간이 지닌 고유의 멋을 창출해야 한다는 뜻이다.

Light is Something You Feel with Your Heart.

Before you start creating light, keep three things in mind about light. First, light that helps people stay safe. It is a safe light that protects people from any danger, anxiety, or sense of crisis that may occur at night. Even when creating a beautiful night, safety should come first. Second, it is to create a light that shows the nature of the night. Artificial light is not the right way to create light. Nighttime is a unique time of darkness, and you need to keep in mind a "caring light" that preserves the night's natural appearance. Third, create a sense of place with understated light. This means that you need to create a sense of place that is unique to the space.

경·景

일정한 공간이나 건축물에 밤의 매력을 담아내기 위해서는 빛을 어떻게 디자인해야 할까? 빛 디자인의 또 다른 요소는 '풍광(風光)'이다. 우리는 일반적으로 아름다운 경치를 경관(景觀)이라고 한다. 경관은 경치를 바라본다는 의미이지만, 오히려 풍광이라는 단어가 더 적절하다고 본다. 왜냐하면 장소 특유의 정서가 내재한 의미로, 시시각각 변화하는 자연과 우리 삶의 모습을 감성으로 느낀다는 의미를 포함하기 때문이다. 특히, 나는 그 안에 우리 고유의 정(情)이 담겨있다고 생각한다.

SOUL OF NIGHTSCAPE

How do you design light to capture the allure of nighttime in a given space or architecture? Another element of light design is 'scenery'. We usually refer to a beautiful view as a scenery. While scenery means to look at a view, I think the word landscape is more appropriate because it has an inherent sense of the emotion of a place, which includes the emotional feeling of the ever-changing nature and our lives. In particular, I think it contains our own emotions.

별이 빛나는 밤 | Starry Night

풍광이란 바람과 빛을 뜻하며, 빛과 바람이 빚어낸다는 뜻

태양에서 내려오는 빛 중에 우리의 시각을 통해 인지되는 가시광선(Visible Light) 영역(380nm~780nm)에는 붉은색을 띠는 장파장(780nm)의 뜨거운 열선과 파장이 짧은 푸른색의 단파장(380nm)이 있다. 우리는 여름철 해변에서 파란 하늘을 보고 뜨거운 빛으로 태닝(Tanning)을 즐긴다. 태양에서 지구로 오면서 푸른색의 단파장은 먼 대기 중 공기 분자 또는 미립자에 의해서 산란되어 파란 하늘을 만들어 주고, 뜨거운 열을 지닌 장파장이 우리에게로 와서 태닝을 할 수 있는 것이다. 붉게 하늘을 물들이는 저녁 노을 빛도 같은 원리이다. 태양 빛이 지평선 가까이 통과하는 동안 파장이 짧은 푸른색의 빛은 공기 분자 또는 미립자에 의해 멀리서 산란되어 관측자가 있는 곳까지 도달하지 못하지만, 파장이 긴 붉은색의 빛은 관측자가 있는 곳까지 도달되기 때문에 열기는 낮아지고 붉게 물든 저녁 노을빛을 볼 수 있는 것이다. 즉, 빛은 공기 중의 미립자에 따라 빛의 색이나 모습이 달라진다는 것이다. 그 미립자의 변화되는 모습을 만들어 내는 요소가 바로 바람이다. 바람은 순간마다 달라진다. 그래서 노을의 색감은 단 하루, 한순간도 똑같은 빛이 없을 만큼 변화무쌍하다.

과거 우리 조상은 아름다운 자연경관을 나타내는 극히 감성적인 말로 풍광을 표현해 왔다. 나는 빛의 풍광을 나타내려고 노력한다. 풍광이라는 말 안에는 자연의 아름다움을 넘어 우리의 정서와 생활문화가 내포되어 있다. 단순히 보이는 것만을 빛으로 나타내는 것이 아니라 공간 속에 변화하는 시간과 사람들 삶의 모습, 그 장소만의 이미지를 그 안에 담으려는 것이다. 매력있는 빛이란 인공적인 조명으로 빚어내는 것이 아니다. 바로 풍광을 담아내는 빛이고, 이러한 빛이 공간에 스며들 때 비로소 감성적인 빛을 느낄 수 있는 것이다. 그러기 위해서는 몇 가지 특성을 고려해야 한다.

먼저, 중요하게 바라보려는 관점을 설정하는 일이다. 우리가 일반적으로 바라보는 방법은 여러 가지가 있다. 첫째, 스카이라인(Skyline) 조망법이다. 이는 아래에서 위를 올려 보는 방법인데, 인공적인 구조물과 자연의 모습을 동시에 인지하는 조망법이다. 둘째, 높은 지점에서 아래로 내려다볼 때이다. 이러한 조망을 파노라믹(Panoramic) 조망이라고 한다. 셋째, 움직임의 관점이다. 일정한 구간을 천천히 걸어가면서 보는 조망과 공간을 빠르게 차를 타고 이동하면서 보는 모습이다. 넷째, 조망 거리 관점이다. 가까이서 볼 때도 멀리서 보는 것처럼, 조망하는 거리에 따라서 대상이 달리 느껴지기 때문이다.

Scenery is a Combination of Wind and Light, Meaning that It is Created by Light and Wind.

In the visible light region (380nm to 780nm) of the light from the sun that is perceived by our eyes, there are hot, long-wavelength (780nm) rays that are reddish in color and shorter wavelength (380nm) rays that are blue in color. We see blue skies at the beach in summer and enjoy tanning from the hot light. On their way to Earth from the sun, the blue short wavelengths are scattered by air molecules or particulates in the distant atmosphere, giving us blue skies, and the hot, long wavelengths come to us, giving us a tan. The same principle is responsible for the reddish color of the evening sunset. As the sun passes near the horizon, the shorter wavelengths of blue light are scattered away by air molecules or particulates and do not reach the observer, but the longer wavelengths of red light do reach the observer, which reduces the heat and gives us the reddish color of the sunset. In other words, light changes color or appearance depending on the particles in the air. Wind is the factor that creates the changing appearance of those particles. The wind changes from moment to moment. As a result, the color of the sunset is never the same from one day to the next.

In the past, our ancestors described landscapes with extremely emotional words to describe beautiful natural scenery. I try to represent landscapes of light. The word "landscape" implies not only natural beauty, but also our emotions and living culture. I do not simply represent what I see with light, but I try to capture the changing time and people's lives in the space, and the image of the place. Attractive light is not created by artificial lighting. It is the light that captures the landscape, and it is only when this light permeates the space that it becomes emotional.

To do this, few characteristics should be considered. It is important to establish several perspectives. First, there's the skyline view. This is a way of looking up from the bottom, which recognizes both man-made structures and natural features. Second, you can look down from a high vantage point. This is called a panoramic view. Third, the perspective of movement. It is the view of walking slowly through a certain area, and the view of moving quickly through a space in a car. Fourth, the perspective of distance. This is because objects are perceived differently depending on the distance from which they are viewed, such as when viewed up close and from a distance.

이런 특성을 고려하여 빛 디자인의 관점을 설정해야 한다. 대상을 바라보는 사람들이 어디에서 어떤 방식으로, 또 무엇을 느끼게 해야 할지의 문제가 중요하다. 이는 대상을 단순히 '보다(See)'라는 피상적인 의미를 넘어, 사람들이 대상을 보고 이해하고, 감성적인 느낌까지 더해지는 '지각한다(Perceive)'라는 관점으로, 빛의 감성을 표현하는 것이 중요하다. 대상에 대한 조망지점이 선정되었다면, 그다음 단계로 이제 빛을 디자인해야 한다. 빛은 주요 조망지점에 따라 각기 다른 빛 디자인 연출이 필요하며, 이에 따라 대상의 성격이 각기 다르게 나타난다.

낮의 도시는 태양 빛으로 그 도시의 독특한 풍광을 나타내기가 어렵지만, 밤에는 어둠이라는 캔버스 위에 자연과 인공의 빛이 더해져서 우리가 표현하고자 하는 밤의 정경을 연출할 수 있다. 감성을 담는 빛에 따라 장소 본연의 독특한 풍광을 만들어 내는 것이다. 매력이 넘치는 풍광을 연출하기 위해서는 구조적인 관점에서 빛의 구성을 볼 수 있어야 한다. 변하지 않는 공간 요소로서, 자연의 빛과 인공의 빛이 함께 어우러지는 모습을 통해서 빛의 이미지를 인지하는 것이다. 이는 대상을 밝힌다거나 여러 가지 색으로 비추어서 보여주는 단순 조명법이 아니다.

이런 관점에서 빛이 다양한 색상을 사용하거나 모두 밝아야 할 필요가 없다. 즉, 빛과 어둠의 조화가 감성적인 밤의 모습을 빚어낸다. 이는 밤이라는 캔버스에 빛을 더해서 공간을 연출하는 것만이 아니라 공간에서 불필요한 인공조명을 빼는 방법으로, 밤의 감성을 고려하여 공간 본연의 모습을 담아내기 위한 다양한 방법이 중요하다. 좋은 빛이란 감성을 담아내는 풍광이라는 관점에서, 먼저 빛으로 어둠을 배려하려 할 때 비로소 공간과 장소에 생명력을 불어넣을 수 있다.

These characteristics should be taken into account when setting the perspective of your light design. It is important to think about where, how, and what the viewer should feel when looking at the object. It is important to go beyond the superficial meaning of 'see' to 'perceive,' which is how people see, understand, and feel the object, and to express the emotion of light. Once the viewpoints for the object have been selected, the next step is to design the light. Each key vantage point requires a different light design, and the character of the object is expressed differently accordingly.

During the day, it is difficult to capture a city's unique landscape with sunlight, but at night, the canvas of darkness is filled with natural and artificial light to create the night scenery we want to portray. The light that captures the emotions of a place creates a unique landscape. In order to create a landscape full of charm, we need to be able to see the composition of light from a structural point of view. As an unchanging spatial element, the image of light is perceived through the way natural and artificial light work together. It is not just about lighting an object or showing it in different colors.

From this perspective, light does not have to be multi-colored or all bright to create an emotional nighttime scene, but rather a harmony of light and darkness. It is not just about adding light to the canvas of night to create a space, it is also about removing unnecessary artificial light from it, and it is important to consider the emotions of the night to capture the essence of the space. Good light is a landscape that captures emotions, and only when we consider the darkness with light can we bring spaces and places to life.

색온도가 낮은 편안한 빛
Comfortable light with low color temperature

색온도가 높은 생동감 있는 빛
Vibrant light with high color temperature

빛의 디자인, 울산광역시 울산교 댄싱 브릿지
Light Design, Ulsan Dancing Bridge, Ulsan Metropolitan City, 2023

빛 연출가가 밤 자체의 속성인 어둠을 소중하게 생각하는 이유는 어둠 속에서 달빛과 별빛을 볼 수 있기 때문이다. 빛 연출가에게 어둠은 빛의 붓으로 많은 것을 그려낼 수 있기 때문에 매력적인 요소가 된다.

Lighting directors value darkness as an attribute of night itself because it allows them to see moonlight and starlight. For a lighting director, darkness is fascinating because so much can be painted with the brush of light.

빛·光

좋은 빛으로 공간에 생명감을 담아낸다

빛은 공간에 새로운 생명을 부여하는 씨앗이다. 밤이라는 캔버스에 빛의 붓으로 공간을 하나씩 그려간다. 밤을 지키고자 하는 절제된 마음의 빛으로 현란하거나 과장되어 보이지 않는, 그 공간과 장소에 적합한 새로운 생명을 담는 감성의 빛 그림을 그리는 것이다. 밤을 만들어 내는 빛은 노을빛, 별빛, 달빛과 어우러진 인공적인 조명, 그리고 이러한 빛 속에서 사람들 삶의 모습이 만들어 내는 빛이 더해질 때 아름다운 공간과 장소가 태어난다.

밤을 존중하는 빛의 디자이너이자 총감독

나는 빛의 디자이너이자 빛 연출 총감독*으로서 빛을 디자인하는 핵심 요소인 밤을 존중한다. 밤 속에서 최소한의 빛으로 그 대상, 그 공간이 담고 있는 장소성을 담아내는 것이 무엇보다도 중요하다. 밤의 안전한 모습에 장소나 공간이 가진 개성과 이미지를 드러낼 빛을 디자인하는 것이다. 곧, 빛의 채움이 아닌 빛의 비움의 마음으로 공간을 바라볼 수 있어야 한다. 바로 절제된 감성의 빛이 필요한 것이다. 우리는 대개 밤을 밝히려고만 한다. 그러나 한번 무절제하게 밝혀진 밤의 공간은 다시 어둠을 지켜나가기에는 많은 어려움이 따른다. 바로 빛이 가진 생명의 마력(Power)이 존재하기 때문이다.

LIGHT

Bringing a Space to Life with Good Light

Light is the seed that gives new life to a space. On the canvas of night, I paint each space with the brush of light. With the light of a restrained mind that wish to preserve the night, I paint a sensitive light painting that does not look flashy or exaggerated, but rather a new life that is appropriate to the space and place. The light that creates the night is a combination of sunset, starlight, moonlight, artificial light, and the light of people's lives in these lights, and when added together, beautiful spaces and places are born.

Designer and General Manager of Light Honoring the Night

As a light designer and director of light production*, I respect night as a key element of light design. It is of utmost importance to capture the object and the sense of place that the space contains with minimal light in the night. It is to design a light that will reveal the personality and image of a place or space in the safety of night. In other words, it is to be able to look at the space with the mind of emptiness of light, not the filling of light. That's what understated light is for. We often try to light up the night, but once a space has been illuminated indiscriminately, it is difficult to keep it dark again. This is because of the life-giving power of light.

---

*빛을 디자인 설계 후, 현장 시공 시 발생하는 빛의 디자인 부분에 대한 감리 개념의 의미를 나타내는 단어이다. 기존 전기공사 감리는 전기공사 부분의 기술적인 부분에 머물던 요소이며, 빛의 컬러(Color)와 연출 방법, 빛을 비추고자 하는 방향과 설치 위치의 확정, 점등의 계획과 밝기 단계를 정하는 사항 등 다양한 빛의 디자인 설계가 현장에서 완성도 있게 만들어질 수 있도록 공사업체와 자재업체를 총괄 관리하는 업무로서 빛 연출의 작업을 완성하는 가장 중요한 부분의 최종 공정을 책임지는 업무이다.

*General supervision of light design: A word that expresses the meaning of the concept of supervision of the light design part that occurs during on-site construction after the design of the light design. Previously, the supervision of electrical work was limited to the technical part of electrical work, and it is the task of managing construction companies and material companies so that various light design designs such as color and production method of light, confirmation of the direction and installation location of light, lighting plan and brightness level, etc. can be completed on site, and is responsible for the final process of the most important part of completing the work of light production.

흥인지문 빛 연출 스케치와 현장 테스트 | Sketch and field-testing, Light Design for Heunginjimun Gate

빛의 디자인, 보물 제1호 서울시 흥인지문(興仁之門) | Light Design, Treasure No. 1: Heunginjimun Gate, Seoul, Korea, 2014

좋은 빛, 자연과 조화에서 찾는다

나의 빛 모토는 자연이다. 나는 자연의 빛에서 진정한 빛의 모습을 느낀다. 나의 빛 연출은 자연이 담고 있는 빛을 조화라는 관점으로 공간과 장소에 빛을 담아내려고 노력하는 것이다. 나는 나를 소개할 때 '조명디자이너'라 하지 않고 '빛의 디자이너'라고 한다. 나는 빛 디자인에서 자연의 빛 개념을 핵심 요소로 사용하기 때문이다.

공간을 디자인하는 빛은 인공조명을 연출하기 위한 조도(照度), 휘도(輝度), 색온도(色溫度) 등의 정량적인 측면을 중심에 두고 생각하지만, 내가 디자인하는 공간의 빛은 먼저 자연이 베풀어 주는 감성의 빛을 담으려고 노력한다.

Finding Good Light in Harmony with Nature

My motto of light is nature. I feel the true form of light in the light of nature. My light production is to try to bring light into spaces and places from the perspective of harmonizing the light contained in nature. When I introduce myself, I do not say 'lighting designer' but 'designer of light'. This is because I use the concept of natural light as a key element in my light design.

While the light in space design is centered on quantitative aspects such as illuminance, luminance, and color temperature to create artificial light, I try to capture the emotional light of nature first.

서울시 청계천 복원건설공사 3공구 빛의 디자인, 국제도시조명연맹(LUCI), CPL Award 1등 수상 | Lighting Urban Community International (LUCI), CPL "City.People. Light" Award, First Prize, Light Design for District 3 of Cheonggyecheon Restoration Project, Seoul, Korea, 2008

좋은 빛, 사람들 삶의 모습을 밝힌다

나는 2003년 서울시 청계천 복원건설공사 당시 청계천 3공구인 청계천 8가에서 신답철교까지 2km 구간의 빛을 총괄 디자인했다. 같은 시기에 한국 제2의 도시 부산광역시로부터 부산의 온천천이라는 생활 하천에 대한 빛 연출 디자인을 요청받았다. 서울시 청계천은 서울의 중심부를 흐르는 생활형 하천으로 주변에는 상업지역과 왕십리 뉴타운의 대규모 주거지역이 위치한 곳이다. 또한 종로, 광교, 동대문시장을 지나 중랑천과 합류되는 곳이기도 하다.

    부산시 온천천은 생활형 하천이지만 청계천보다 넓은 하천이었다. 주변은 주거밀집 지역으로, 고층아파트가 위치한 특성을 보이고 있었다. 사실상 청계천처럼 빛을 설계할 수 있는 장소가 아니었다. 다만, 청계천처럼 시민들이 자주 찾는 공간으로 형성되어 있어서 빛이 온천천의 공간을 더욱 생동감 있게 만들어 낼 수 있었다고 생각했다. 서울시 청계천의 일부 공간은 관광객이 찾는 공간이고, 나머지 공간은 수변을 따라 형성된 주민들의 산책 공간의 빛이 필요했다. 이 빛은 저녁 노을 같이 편안함이 담긴 빛을 만들어 냈다.

Good Light, Illuminating People's Lives

In 2003, during the restoration and construction of Cheonggyecheon Stream in Seoul, Korea, I was responsible for the overall design of a 2-kilometer section of light from Cheonggyecheon 8th Street in Cheonggyecheon District 3 to Sindap Railroad Bridge. Around the same time, I was asked by Busan Metropolitan City, Korea's second largest city, to design a light production for a living stream called Oncheoncheon in Busan. Seoul's Cheonggyecheon is a living stream that flows through the center of Seoul, with commercial areas and a large residential area in Wangsimni. It also flows through Jongno, Gwanggyo, and Dongdaemun markets before joining Jungnangcheon.

    Oncheoncheon Stream in Busan is also a living stream, but it is wider than Cheonggyecheon Stream. The surrounding area was densely populated and characterized by high-rise apartments. In fact, it was not a place where light could be designed like Cheonggyecheon. However, like Cheonggyecheon, it was formed as a space frequented by citizens, so we thought that light could make the space of Oncheoncheon more lively. In Seoul's Cheonggyecheon, some of the space is for tourists, while the rest of the space needed to be illuminated by the light of the residents' walking space along the water's edge. This light creates a comforting glow like an evening sunset.

서울시 청계천 비우당교와 두물다리 빛의 디자인 | Light Design of Biwoodang Bridge and Duemul Bridge, Cheonggyecheon, Seoul, 2005

일출 | Sunrise

일출의 장관(壯觀). 눈부시게 아름다운 자연의 빛이다.
빛의 기본은 자연이다.

The spectacle of a sunrise. It is a dazzling display of natural light.
The basis of light is nature.

Chapter II
빛을 느끼다:
景

SENSING THE LIGHT: SCENERY

빛에 따라 공간의 이미지가 변화한다
우리는 빛을 느끼기보다는 단순히 본다는 관점에서 공간과 사물을 이해하려는 경향이 있다. 빛이 없다면 우리는 공간의 깊이, 형태, 색과 이미지 등을 보고 느낄 수가 없다. 빛을 느끼기 때문에 우리의 시각에서 공간의 특성을 인지할 수 있는 것이다. 빛을 보고 인지하는 방법에 따라 밤 공간이 달라지기 때문에 빛 연출가는 빛을 다양하게 연출한다.

The Image of Space Changes Depending on the Light
We tend to understand space and objects in terms of what we see rather than what we feel. Without light, we cannot see and feel the depth, shape, color, and image of space. It is because we feel light that we can perceive the characteristics of space from our perspective. Because the way we see and perceive light changes the way we perceive a nighttime space, light designers create different types of light.

빛 연출 정의

DEFINITION OF LIGHT PRODUCTION

빛 연출이란 밤이라는 시간과 공간에 절제된 빛과 다양한 요소로, 생명감이 감도는 감성의 공간을 만들어 내는 것을 말한다. 매력 있는 빛 연출을 통해 많은 사람에게 삶의 행복을 선사해 주는 것이다. 나는 빛을 디자인할 때 항상 "밤을 배려한다."라는 말을 자주 사용한다. 밤의 공간에 빛이 서로 경쟁하지 않고 서로 조화있게 어울릴 때 빛의 가치가 살아있는 장소성을 담아내기 때문이다.

Light production is the use of understated light and various elements in the time and space of night to create an emotional space full of life. By creating attractive light, we can bring happiness to many people's lives. I often use the phrase "consider the light" when designing light. The value of light in a nighttime space is when it harmonizes with each other instead of competing with each other, creating a sense of place.

빛 연출 대상

밤에 보이는 대상은 크게 산과 강, 공원과 호수 등의 자연적 요소와 도시와 건축물 등의 인공적 요소로 나뉜다. 밤의 모습을 드러내는 빛은 주로 문화적 요소인 도시와 건축물 등 사람이 인지하는 공간에 연출한다. 자연적 공간인 숲이나 사람들이 잘 찾지 않는 자연경관에 더해진 빛은 실용적인 측면에서 전기세만 낭비하고, 자연 본연의 모습을 훼손하는 요소가 될 것이기 때문이다. 만약 빛으로 공간을 밝혀야 한다면 안전을 보장해 줄 수 있는 최소한의 빛으로 주말이나 특정일의 특정 시간에만 빛이 디자인돼야 한다. 나머지 시간은 자연에 그대로 돌려줘야 한다. 자연의 아름다운 풍광을 빚어내는 달빛, 별빛, 노을 등 자연의 빛만으로도 매력이 넘치는 모습을 연출하고 있지 않은가. 빛 연출 대상에 자연의 빛을 고려하면서 밤의 풍광을 만들어 가려는 빛 연출가의 마음가짐이 중요하다.

빛 연출은 현대적 개념이며, 절제된 빛이어야 한다.

10년 혹은 20년 전의 원색적인 조명연출은 요즘 사람들의 기대치나 눈높이와 크게 다른 것이었다. 요즘은 안전을 갖추는 것은 물론이고 가족과 함께할 수 있는 행복한 밤의 시간과 공간에 방문객이 모여들고, 사람과 사람이 만나는 행복한 문화공간이 창출되는 것이다. 곧, 연인과 가족이 함께 가고 싶은 감성이 담긴 편안한 공간이어야 한다는 것이다.

빛은 사람들이 편안하게 다가올 수 있는 아늑한 공간을 만드는 데 중요한 요소이다. 그렇다고 빛 자체가 주연은 아니다. 공간을 방문하는 사람들이 주인공이면서 그 속에서 빚어내는 행복한 사람들의 모습이 자연스럽게 어우러질 수 있는 빛이어야 하고, 이때 빛은 있는 듯 없는 듯해야 한다.

OBJECT OF LIGHT PRODUCTION

The objects visible at night are divided into two main categories: natural elements, such as mountains, rivers, parks, and lakes, and man-made elements, such as cities and buildings. The light that reveals the night is mainly created in spaces that people recognize, such as cities and buildings, which are cultural elements. In the case of natural spaces, such as forests and lesser-visited natural landscapes, light is not only a waste of electricity, but also a detriment to the natural environment. If you need to illuminate a space with light, it should be designed for specific times of the day or weekends, with the minimum amount of light to ensure safety. The rest of the time, it should be left to nature. Moonlight, starlight, sunsets, and other natural light that creates beautiful landscapes are mesmerizing. It is important that the light designer's mindset is to create a night scene while taking into account the natural light.

Light is a Modern Concept, and It should be Understated.

The primary-colored light production of 10 or 20 years ago was very different from the expectations of the people today. Nowadays, it is not only about safety, but also about creating a happy nighttime and space for families, attracting visitors, and creating a happy cultural space where people meet people. In other words, it is about creating a comfortable space with a feeling that couples and families want to go together.

Light is an important element in creating a cozy space that people feel comfortable approaching. However, it is not the star of the show. The people who visit the space are the main characters, and the light should blend in naturally with the happy people it creates, and the light should be both present and absent.

빛의 마스터 플랜, 서울시 야간경관 기본계획 | Light Master Plan of Light Pollution Prevention Plan, Seoul, 2015

도시의 인공구조물과 건축물, 교량, 공원 등에 인공조명 연출을 통해 도시공간의 밤을 아름답게 조성하는 것도 중요하지만, 그 전에 사람들이 편안하게 머물 수 있는 빛의 공간을 먼저 만들어야 한다. 아직도 많은 사람이 인공적인 조명이 단순히 공간을 밝게만 밝혀주기만 하면 된다고 생각하고, 눈부시게 밝거나 형형색색의 화려한 조명만을 만들어 내고 있다.

While it is important to beautify urban spaces at night by creating artificial light on man-made structures, buildings, bridges, parks, and more, it is also important to create light spaces that people can feel comfortable in. Many people still think that artificial lighting is just about brightening a space, creating blindingly bright, colorful, or flashy lights.

빛의 디자인, 소노호텔&리조트 쏠비치 진도 | Light Design, Sono Hotels & Resorts, Sol Beach Jindo, 2019

빛 연출 이미지

우리는 해가 떠서 해가 지는 하루 동안 다양한 빛의 색상이나 밝기에 따라서 각기 다른 감성을 접한다. 한낮의 맑고 푸른 하늘빛에서 청명함을 느끼고, 붉게 물든 노을에서 편안함을 만난다. 이렇게, 청명한 빛은 밝고 활기 있는 공간 이미지를 만들고, 석양은 평온과 행복의 공간 이미지를 만든다.

우리가 보는 햇빛은 살아있다. 일출에서 일몰까지 순간순간 변하여 자연의 빛은 결코 멈춰진 빛이 없다. 우리는 늘 변화하는 빛, 살아 움직이는 빛의 변화에 따라서 새로운 공간과 장소를 보게 되는 것이다. 우리에게 인지되는 빛은 크게 두 가지 변화를 통해 만나볼 수 있는데, 첫째 빛의 위치이다. 빛의 위치에 따라서 빛의 밝기가 변화한다. 둘째 빛의 색감으로, 느낌이다. 빛의 색감에 따라 느끼는 감성이 달라진다. 빛이 변화하는 모습은 공간에 다양한 이미지를 창출해 내는 중요한 요소인데, 나는 빛의 다채로운 이미지를 여섯 가지로 구분했다.

IMAGES OF LIGHT PRODUCTION

As the sun rises and sets throughout the day, we experience different emotions depending on the color or brightness of the light. We find clarity in the clear blue sky at midday and comfort in the ruddy sunset. In the same way, clear light creates a bright and energetic spatial image, while a sunset creates a spatial image of calm and happiness.

The sunlight we see is alive, changing from sunrise to sunset and consistently shaping the world around us. This ever-changing light creates new perspectives and spaces in nature. The light we perceive undergoes two primary changes: its position, which affects its brightness, and its color, which impacts the emotions it evokes. The varying appearance of light plays a vital role in shaping different visual experiences, leading me to categorize the colorful pictures of light into six categories.

일출에서 일몰까지 빛의 변화되는 모습 | The changing appearance of light from sunrise to sunset

부드럽고 따스한 빛

온기(溫氣)의 빛은 밝기가 낮다. 그래도 빛의 색감은 따뜻하다. 빛의 방향이 어느 한쪽에서 특별하게 나타나지 않는다는 것이 특징이다. 그래서 그림자가 명확하지 않다. 하늘에 빛나는 부드럽고 섬세한 구름의 따뜻한 색감은 우리에게 평온함을 선사하는 것과 같다. 실내 공간에서 온화한 느낌을 원할 때는 바닥으로 비추는 직접적인 다운 라이트(Down Light) 조명을 사용하지 않고 상부로 비추는 따스한 간접적인 업라이트(Up Light) 조명을 사용하면 실내 공간이 부드럽고 평온하게 인지된다. 편안한 느낌의 공간은 이렇게 빛의 방향성과 색감으로 만들어진다.

Soft and Warm Light

Soft and warm energy has a lower brightness, but the color of the light is still warm. It is characterized by the fact that the direction of the light is not particularly dominant on either side. As a result, shadows are not obvious. It is like the warm colors of soft, delicate clouds in the sky that give us a sense of calm. When we want to create a warm feeling in an internal space, we should not use direct down light that shines down to the floor, but use warm indirect up light that shines upwards, and the internal space is perceived as soft and calm. This light direction and color scheme creates a relaxing space.

맑고 시원한 빛

청기(淸氣)의 빛은 밝고 시원하다. 공간을 명확하고 선명하게 인지시키는 빛이다. 여름날 정오에 파란 하늘빛과 푸른 들녘을 밝고 선명하게 비추어 주는 빛으로, 방향성은 위에서 아래로 수직적이고, 빛의 색감은 맑고 선명한 청백색의 빛으로서 명확한 그림자를 만들어 내는 밝은 빛이다. 청명한 하늘 아래 탁 트인 대지를 바라보는 것과 같이 빛에 시원한 기운이 담겨있다. 우리의 생활 속에서 청기의 빛은 밤에 축구장과 대형 스타디움에 설치된 조명법이라고 생각하면 쉽게 이해할 수 있다. 한낮의 모습처럼 활동적인 공간으로 디자인하고 싶다면 높은 폴(Pole) 타입의 조명을 사용하여 청백색의 빛으로 밝게 연출하면 맑고 시원한 이미지를 연출할 수 있다.

Clear and Cool Light

The light of cheonggi(淸氣) is bright and cool. It is a light that gives a clear and sharp perception of space. It is the light that brightly and clearly illuminates the blue sky and green fields at noon on a summer day. The direction is vertical from top to bottom, and the color of the light is clear and crisp blue-white light that creates clear shadows. There is a coolness to the light, like looking at the open land under a clear sky. In our daily life, blue light can be easily understood by thinking of the lighting system installed on soccer fields and large stadiums at night. If you want to design an active space that looks like it does in the middle of the day, you can use high pole-type lights to brighten it with blue-white light to create a clear and cool image.

생동감을 주는 빛

생기(生氣)의 빛은 편안한 밝기로, 빛의 색감은 파스텔 색조의 부드러운 기운을 담고 있다. 생기있는 빛은 오로라처럼 하늘에서 자연 현상으로 만들어진 또 다른 색감이 담긴 빛으로, 주변의 공간을 화려한 빛으로 물들인다. 이런 빛의 속성을 인공조명으로 연출할 때는 화려한 불꽃 축제나 레이저 쇼 등의 빛으로 연출하여 공간을 화려하게 만든다. 인공적인 빛의 요소를 활용하여 자연에서 볼 수 있는 다양한 색감을 나타낸 것이다. 이러한 생기있는 빛은 사람들을 행복하게 만든다. 그러나 모든 빛이 화려하다면 진정으로 아름답지 않을 수 있다. 화려한 빛은 담백한 빛과 함께 연출될 때 강렬하고 아름답게 느껴진다.

Vibrant Light

Light of vitality is a relaxing brightness, and the color of the light is soft with pastel shades. Vibrant light is another colorful light that is created by natural phenomena in the sky, such as the aurora, which bathes the surrounding space in colorful light. When these properties of light are created with artificial light, they can be used to create a spectacular fireworks display, laser show, or other light show to make a space pop. The artificial element of light is used to represent the different colors found in nature. This vibrant light makes people happy, but if all light is colorful, it may not be truly beautiful. Colorful light feels intense and beautiful when it is paired with subdued light.

마음을 담은 빛

동기(動氣)의 빛, 빛의 밝기는 낮으나 빛의 색감은 단순하다. 빛의 볼륨은 크지 않다. 감성의 작은 빛 하나로도 공간이 가진 이야기를 충분히 나타낼 수 있게 해준다. 빛의 형태가 작을수록 더욱 감성적으로 변화한다. 밤하늘에 작게 반짝이는 별빛, 아주 작은 바람에도 빛이 흔들리는 촛불의 움직임처럼 차분한 빛이다. 감성을 통해 자연에서 느껴지는 숭고함과 역사 문화를 담고 있는 요소들의 고즈넉함을 연출할 수 있는 빛이다.

Light Filled with Emotion

The motivational light is not very bright, but it has a simple color. It is not very strong in volume. A small, emotional light is enough to convey the story of the space. The smaller the light's shape, the more emotional it becomes. It is a calm light, like a small twinkling star in the night sky or the movement of a candle whose light sways with the slightest breeze. This light can evoke the grandeur of nature through emotion and the quietness of elements that contain history and culture.

살아있는 빛

정기(精氣)의 빛은 밝기가 강하다. 빛의 색감도 강렬하다. 살아 움직이는 빛은 하늘과 대지를 가로지르는 낙뢰(落雷), 천둥 번개가 칠 때 번쩍거리는 울림과 순간적인 폭발과 사라짐에서 살아있는 빛을 만날 수 있다. 무대 공간에 인공적인 조명을 사용하여 연출할 때 섬광의 효과를 주는 빛과 흔들리는 빛을 극적으로 만난다. 활활 타오르는 모닥불의 모습은 빛과 열이라는 두 가지 요소를 직접 느끼게 해주는 살아있는 빛이다.

Energetic Light

The bright and vibrant energy of life emits intense brightness and vivid colors. This living light is visible in the lightning streaks across the sky and earth, in the flashing bursts of thunder, and in the sudden bursts and disappearances. When artificial light is used to illuminate a stage, it can be combined dramatically with flickering and wavering light. A roaring bonfire embodies living light, providing a direct experience of both light and heat.

雲

성스러운 빛

운기(雲氣)의 빛은 선구자의 카리스마처럼 힘차다. 빛의 색감은 무채색이다. 구름 사이를 뚫고 쏟아지는 밝은 빛, 그리고 방향성을 줄 만큼 강렬한 빛, 교회와 사찰, 성당과 같은 종교 시설 공간에서 이런 성스러운 빛을 만날 수 있다. 성스러움의 요소는 우리가 일상에서 경험하지 못하는 신비감을 주는 빛이다. 서양 중세기 성당에서 흔히 볼 수 있는 커다란 장미창(Rose Window)에 만들어진 스테인드글라스, 이곳을 통해 보이는 빛의 모습은 천상의 신비하고 성스러운 빛을 창출해 낸다.

Sacred Light

The light of the sacred energy is powerful, like the charisma of a harbinger. The color of the light is achromatic. You can find this sacred light in the spaces of religious institutions, such as churches, temples, and cathedrals, where the light is bright, breaking through the clouds, and intense enough to give direction. The element of sacredness is light that gives us a sense of mystery that we do not experience in our everyday lives. Stained glass in a large rose window, common in medieval cathedrals in the West, creates a celestial, mysterious, and sacred glow.

빛으로 공간 이미지 디자인

빛이 공간의 이미지를 어떻게 창출해 낼 수 있는지에 대해서 몇 건축가의 작품을 통해 이야기하고자 한다. 먼저, 일본의 건축가 안도 다다오(安藤忠雄, 1941~)의 건축공간이다. 우리는 삶에 지쳤을 때 위안을 얻기 위해 종교를 찾고 의지한다. 그래서 종교건축에서는 영혼이 지친 사람을 위무할 평온한 공간을 위해 다양한 관점에서 배려하고 기획한다. 교회 건축물의 큰 십자가는 멀리서도 우리를 위해 따뜻하게 밝혀주고 있다. 교회 내부 공간에서도 십자가의 모습을 볼 수 있다. 그러나 안도 다다오 빛의 교회(1989)에서는 전혀 다른 빛을 만난다. 큰 벽에 위치한 십자가의 모습이 마치 벽의 창문처럼 외부와 내부를 연결하고 있다. 외부에서 들어오는 따뜻한 햇살 속에서 또 하나의 십자가가 내부에 내려와 있다. 바로 바닥에 드리워진 십자가의 모습이다. 빛의 십자가는 우리가 앉아있는 바닥을 비춘다. 거룩하게 여기는 십자가를 가장 낮은 곳에 내려놓고 우리를 반기는 것이다.

Designing Spatial Imagery with Light

Let's explore how light can create the image of a space through the work of a few architects. First, let's focus on an architectural space created by Japanese architect Ando Tadao(1941~). When we feel weary, we often seek solace in religion. This is why, in religious architecture, we carefully consider and plan for a serene space that provides comfort for the soul. In church architecture, the large crosses are warmly illuminated from a distance, and we also see them within the internal space of the church. However, in Ando Tadao's Church of the Light (1989), we encounter a different approach to light. The cross on the large wall acts as a connection between the outside and the inside, almost like a window. The warm sunlight coming in from the outside creates another cross within the interior space, illuminating the floor where we sit. The holy cross is positioned at the lowest level, welcoming us in.

「빛의 교회」, 안도 다다오(安藤忠雄)
Church of the Light, Ando Tadao, 1989

적절한 시간에, 적절한 양의, 적절한 빛을 디자인한다. 「롱샹 성당」르 코르뷔지에
Design of Light at the right time, the right amount, the right light. Ronchamp Chapel by Le Corbusier, 1950-1954

절제된 빛은 성스럽고 아름답다

서울 명동에 있는 명동성당은 종교적인 관점에서 평화를 지향하는 이상향을 잘 나타내고 있다. 우리가 살아가는 현실이 힘들고 암울할 수 있지만 성당의 큰 문을 열고 명동성당의 실내로 들어서면 멀리 보이는 스테인드글라스의 신비로움이 이곳을 방문하는 사람들에게 안식과 마음의 평화를 전해준다. 바로 빛으로 사랑과 안식과 평화를 전해주는 것이다.

Understated Light is Sacred and Beautiful

Myeongdong Cathedral in Seoul, Korea, represents the ideal of peace from a religious perspective. The reality of our lives may be hard and bleak, but when you open the large doors of the cathedral and enter inside, the mystery of the stained glass windows in the distance conveys a sense of rest and peace of mind to those who visit. It is the light that conveys love, rest, and peace.

성스러움을 밝혀주는 빛 | Light that illuminates the sacred

자연의 채광이 더욱 성스럽다

우리가 흔히 알고 있는 근대건축에서 건축의 장식을 철저하게 배제한 건축양식이 있다. 이러한 건축양식 중에서 종교건축은 장식을 최소화하면서 공간이 담고 있는 마음을 나타내야 했다. 근대건축의 아버지라 불리는 르 코르뷔지에(Le Corbusier, 1887~1965)의 작품인 롱샹 성당(Ronchamp Chapel)은 건축적 장식을 전혀 사용하지 않고 빛으로 공간의 깊은 울림을 담아내고 있다. 롱샹 성당의 자연 채광연출은 창문의 깊이, 크기, 경사각 등 자연광이 실내로 들어오는 방식을 각기 다르게 표현함으로써, 빛 자체가 가진 중요한 속성을 건축에 적용하여 장식이 전혀 없는 건축물에 이상세계의 분위기를 담아냈다. 순수한 자연 채광이 롱샹 성당을 더욱 성스럽게 밝혀주는 것이다.

Natural Light is more Sacred

In modern architecture as we know it, there are some architectural styles that completely exclude ornamentation. Among these architectural styles, religious architecture had to minimize decoration and express the spirit of the space. The Ronchamp Chapel, a work by Le Corbusier (1887-1965), the father of modern architecture, uses no architectural ornamentation at all, but instead uses light to capture the deep resonance of the space. By varying the way natural light enters the room through the depth, size, and slope of the windows, Ronchamp Chapel's natural daylighting brings the important properties of light itself into architecture, creating an otherworldly atmosphere in an otherwise unadorned building. Pure natural light illuminates Ronchamp Chapel, making it even more sacred.

우리가 자연에서 빛을 찾고자 하는 이유는 의식적이든 무의식적이든 실생활에서 직접 보고 느끼는 익숙한 부분이 있기 때문이다. 우리의 주변 환경에 대한 분석은 결국 자연과 조화로운 빛 디자인으로 귀결된다. 조화가 잘 된 빛은 감성을 담은 부드러운 풍광을 만들어 낸다. 자연의 빛은 경쟁하지 않고 서로를 배려하여 우리에게 평온한 빛의 공간을 만들어 주기 때문이다.

The reason we look for light in nature is that, consciously or unconsciously, there is something familiar about what we see and feel in real life. Analyzing our surroundings boils down to designing light that is in harmony with nature. Well-harmonized light creates soft landscapes with emotion. This is because the lights in nature do not compete with each other, but rather support each other, creating a calm light space for us.

운무에 깃든 아침 햇살 | Morning Sunlight Nested in the Fog

빛 연출 원칙

빛 연출은 기본적인 원칙을 지킬 때 공간 본연의 가치가 빛날 수 있다. 이는 빛 연출에 대한 나의 신념이다. 첫째, 사람들을 위한 안전한 빛(Safety)이 되어야 한다. 공간을 차갑고 무섭지 않은 따뜻하고 밝은 공간을 만드는 빛이어야 한다. 둘째, 빛 공해가 없는 최소한의 절제된 빛으로 많은 사람에게 아름답고 매력적인 (Beauty) 공간을 만들어 주어야 한다. 셋째, 그 공간만의, 그 장소다운 사람들 삶의 이야기(Identity)가 담겨야 한다.

PRINCIPLES OF LIGHT PRODUCTION

When you follow the basic principles of light design, the intrinsic value of a space can shine through. These are my beliefs about light production. First, it should be a safe light for people. It should be light that makes a space feel warm and bright, not cold and scary. Second, it should be a minimal, understated light with no light pollution that makes the space beautiful and attractive to many people. Third, it should contain the stories of people's lives that are unique to that space and that place.

빛의 디자인, 강원도 횡성군 횡성성당 | Light Design, Hwengseong Cathedral, Hwengseong-gun, Gangwon Province, 2021

따뜻한 빛, 평온함을 느낀다

먼저 안전이 확보된 밤을 조성하는 것이 최우선이다. 보행로와 공원 등 주로 많은 사람이 이용하는 공간은 안전을 확보하여 범죄를 예방하는 것이 빛 연출의 첫째 중요한 원칙이다. 공원에 설치되어 있는 높이 4m의 확산형 보안등은 태양의 모습을 형상화해서 밤의 어둠으로부터 공원 이용자의 안전을 확보하기 위해 설치된 인공조명 시설이다.

Warm Light Feels Peaceful, 「Safety」

Creating a safe environment at night is our top priority. The first important principle of lighting is to prevent crime by ensuring safety in spaces that are frequently used by many people, such as sidewalks and parks. The 4-meter-high diffused security light installed in the park is an artificial lighting facility designed to resemble sunlight and ensure the safety of park users during the night.

안전성

안전을 위한 조명으로서 눈부심 없는 쾌적한 밤의 풍광을 조성한다.

▷ 빛은 안전을 확보하는 데 도움을 주어야 한다.
 - KS A 3011권장 조도를 기초로 안전에 필요한 조도를 확보한다.
▷ 이용자와 조망자에게 시각적 편안함을 주어야 한다.
 - 걸을 때와 운전할 때 빛에 의한 눈부심이 발생하지 않도록 한다.
▷ 공간별 이용자의 성격을 고려한 시간 순응으로 빛의 차별화 된 계획을 연출한다.
 - 빛은 공간과 공간 주변의 성격을 고려하여 빛 연출 방법을 계획한다.
 - 주변의 조도가 낮을 때는 낮은 조도로 시각의 순응을 확보한다.

Safety

Light for safety creates a pleasant nighttime landscape without glare.

▷ Light should help to secure safety.
 - Secure the illuminance required for safety based on the recommended illuminance of KS A 3011.
▷ Provide visual comfort for users and viewers.
 - Avoid light glare when walking and driving.
▷ Create a differentiated light plan with time compliance that considers the nature of users by space.
 - Consider the nature of the space and its surroundings when planning how to produce light.
 - When the surrounding illumination is low, secure visual compliance with low illumination.

빛의 디자인, 제주도 서귀포시 걸매생태공원 | Light Design, Gullmae Ecological Park, Seogwipo-si, Jeju-do, 2021

부드러운 빛, 아름다움을 밝힌다

어두운 밤을 최소한의 절제된 빛으로 아름답고 매력적(Beauty)인 장소로 창출해 내야 한다. 우리는 밤을 더 효율적으로 이용하기 위해 어둠을 밝힌다. 그렇지만 밤 자체의 어둠, 밤의 어둠을 지켜가면서 주변의 환경과 조화있는 연출을 지향해야 한다. 빛의 디자인은 도시를 빛으로 표현하고, 절제된 빛으로 그 도시 그 공간만의 이야기를 담아내야 한다. 빛의 배려와 절제가 없으면 빛은 서로의 경쟁 속에 어느 것도 드러낼 수 없는 무의미한 존재로서 빛의 가치를 잃어버리게 되어 결국 빛이 공해가 될 수 있다.

Soft Light Illuminates Elegance, 'Beauty'

We need to make the dark night a beautiful and inviting place with minimal and understated light. We light up the darkness to utilize the night more efficiently, but we must respect the darkness of the night itself, the darkness of the night, and work in harmony with its surroundings. The design of light should express a city in light, and tell the story of that city and its spaces in an understated light. Without consideration and restraint, light can lose its value as a meaningless entity that cannot reveal anything in competition with each other, and eventually light can become pollution.

심미성
과하지 않는 절제 된 빛의 미학으로 밤하늘의 별빛을 지킨다.
▷ 밤을 위한 빛은 기본적으로 절제미를 지켜야 한다.
 - 랜드마크(Landmark)와 일부 주요 빛 연출 대상에 대해서 점등과 소등계획을 함께 수립한다.
▷ 주간의 모습과 야간의 모습이 서로 조화를 이루어야 한다.
 - 조명시설은 은폐하여 주간에 인지되지 않도록 한다.
▷ 과도한 조명에 의해 생태계에 영향을 발생시키지 않도록 한다.
 - 새어 나오는 빛을 차단하기 위한 빛의 차단 장치와 빛의 밝기 조절 장치 등을 설치하여 빛 공해를 예방한다.

Aesthetics
Protect the starlight in the night sky with the aesthetics of understated light that is not excessive.
▷ Light for the night should be understated in nature.
 - Establish a lighting and extinguishing plan for landmarks and some major light production objects.
▷ The daytime and nighttime look should be in harmony with each other.
 - Lighting fixtures should be concealed so that they are not recognized during the day.
▷ Avoid causing ecological impacts by excessive lighting.
 - Prevent light pollution by installing light shields to block light leakage and light brightness control devices.

빛의 디자인, 안동시 월영교 | Light Design, Wolyeonggyo Bridge, Andong City, 2024

그곳만의 빛, 이야기를 담는다

창의적인 빛으로 그 공간과 그 장소만의 의미와 가치를 담아야 한다. 고유의 장소와 공간만의 이야기가 담긴 독창성과 정체성(Identity)을 담은 빛의 풍광(風光)을 빚어내야 한다. 이는 '이곳만의', '이곳다운' 공간 창출이다. 낮과 다른 밤의 모습을 만들 수도 있고, 낮보다 밤이 더욱 매력이 있는 분위기의 공간이 될 수 있을 것이다.

Unique Light Tells the Story, 「Identity」

The creative light should capture the meaning and value of the space and its location. It is to create a light landscape that contains the originality and identity of a unique place and space with its own story. It is about creating a space that's 'unique' and 'like this place'. You can create a different look at night than during the day, or you can create a space that is more attractive at night than during the day.

장소성

공간의 특성 및 지역의 자연과 역사, 문화를 담아낼 빛의 계획을 수립한다.
▷ 지역의 특화된 개성을 담아내야 한다.
  - 지역의 자연과 역사, 문화를 드러내고 장소만의 성격을 부각시킨다.
▷ 빛으로 시간의 변화를 나타낸다.
  - 시간의 흐름과 공간의 변화를 빛으로 연출한다.
▷ 주변 환경과의 조화를 고려하여 계획해야 한다.
  - 덧셈과 뺄셈으로 빛 연출을 디자인한다.

Sense of place

Establish a light plan that captures the character of the space and the nature, history, and culture of the area.
▷ Capture the specialized personality of the region.
  - Reveal the nature, history, and culture of the area and emphasize the unique character of the place.
▷ Represent the passage of time with light.
  - The passage of time and the change of space should be represented by light.
▷ It should be planned in harmony with the surrounding environment.
  - Design the light production by adding and subtracting.

빛의 디자인, 경상북도 구미시, 김수근 건축가의 구미문화예술회관
Light Design, Gumi Cultural Art Center by Architect Kim Swoo Geun, Gumi, Gyeongsangbuk-do, 2017

세 번째 빛

첫 번째 빛은, 자연 그대로의 빛입니다.
두 번째 빛은, 어둠을 밝히기 위한 인간의 빛입니다.
세 번째 빛은, 사람과 환경을 배려하는 따뜻한 감성의 가치를 담아내는 빛입니다.

The Third Light

The first light is the light of nature.
The second light is the light of humanity to illuminate the darkness.
The third light is the light of warm emotional values that care for people and the environment.

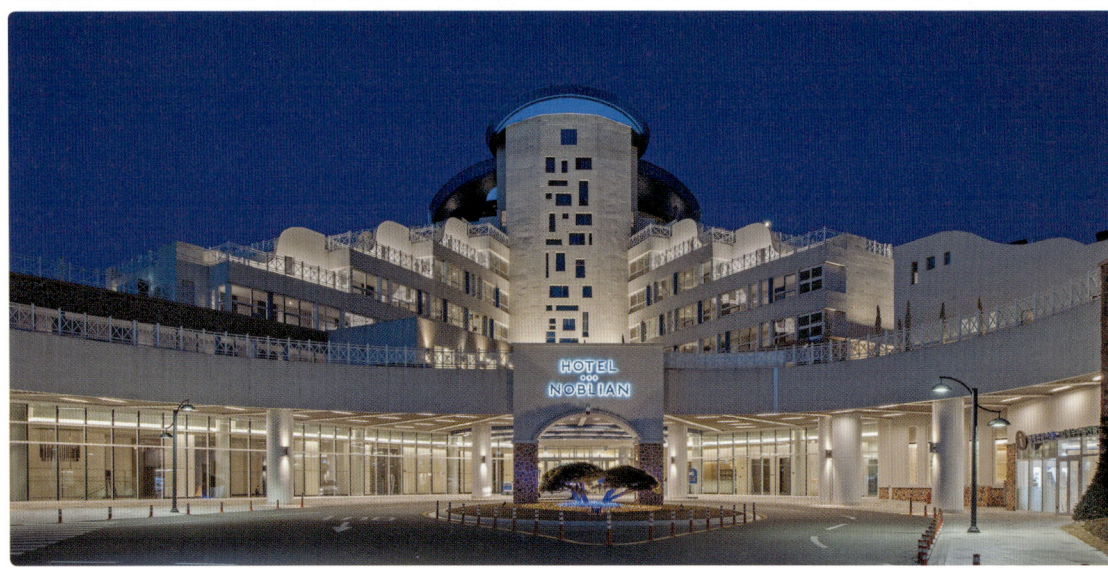

빛의 디자인, 소노호텔&리조트 쏠비치 삼척 | Light Design, Sono Hotels & Resorts, Sol Beach Samcheok, 2015

빛의 디자인, 소노호텔&리조트 쏠비치 진도 | Light Design, Sono Hotels & Resorts, Sol Beach Jindo, 2019

도시 공간은 공연 무대와 다른 다차원 개념이기 때문에 특정 부분에 빛을 연출해서 밤의 이미지를 구축하기란 쉽지 않다. 그리고 도시에는 이미 다양한 빛들이 존재하고 있다. 가로등, 공원에 있는 등, 자동차 불빛, 광고물, 아파트나 저층 주거지에서 새어 나오는 빛 등… 실내 공간이나 무대조명 연출과 다르게, 이미 여러 가지의 빛이 무질서하게 존재한다.

빛 연출에서, 이 모든 빛이 주인공이 될 수는 없다. 주연배우를 밝혀주는 빛은 조연배우를 비추어 주는 빛보다 더 밝게 나타나야 그 특징을 드러낼 수 있다. 어떤 경우는 오히려 빛을 빼는 것으로 주연을 드러내는 방법도 있다. 빛에서도 주연이 있으면 조연이 있어야 한다. 이러한 빛의 위계와 질서의 조화가 공간 무대를 의미 있게 창출해 낸다.

도시를 구성하는 요소인 도로, 보행로, 건축물과 교량 등에서 부각시킬 대상이 무엇인지. 어느 부분이 주연인지를 파악한 다음에 빛을 디자인해야 한다. 중요한 부분은 드러내고, 나머지는 어떤 방법으로 빛을 덜어내거나 낮추어야 할지 고민해야 한다. 빛은 '더 하는 게 아니라 위계와 질서를 부여하는 것이고, 그 조화는 덧셈이 되기도 하고 뺄셈이 되기도 한다.' 라는 것이다. 이는 빛 연출에서 중요한 부분이다.

Because urban space is a multi-dimensional concept, unlike a performance stage, it is not easy to create a nighttime image by directing light to specific areas. And there is already a lot of light in the city. Streetlights, park lights, car lights, advertisements, light leaking from apartments and low-rise residences, etc... Unlike indoor spaces or stage lighting, there is already a chaotic variety of light.

In light composition, all the light cannot be the main character. The light that illuminates the main actor needs to appear brighter than the light that illuminates the supporting actors to reveal their character. In some cases, you can reveal the main character by subtracting light. Even in light, if there is a main character, there must be a supporting cast. This harmony of light hierarchy and order creates a meaningful spatial stage.

Consider the elements that make up your city: the roads, walkways, buildings, bridges, and so on. Figure out which parts are the main elements and then design your light. You need to think about how to bring out what is important, and how to diminish or tone down the rest. Light is not about adding, it is about giving hierarchy and order, and that harmony can be both addition and subtraction. This is an important part of light production.

빛의 디자인, 거제시 신거제대교 | Light Design, Shingeojegyo Bridge, Geoje City, Gyeonsangnam-do, 2016

꺼지지 않는 빛의 도시, 홍콩

홍콩의 밤은 빛의 심포니(Symphony of Light)로 화려한 빛의 도시라는 명성을 얻고 있는 도시이다. 빅토리아 항구(Victoria Harbor)를 중심으로 하버 조명계획(Harbour Lighting Plan)을 적용하기 전에는 각종 건축물 간 경쟁적인 빛 연출로 화려한 빛을 넘어 빛 공해가 극심했던 문제의 도시였다.

그러나 빛의 계획을 적용한 이후에는 스타의 거리(Avenue of Stars)와 홍콩 문화 센터, 완차이 골든 바우히니아 광장(Wan Chai, Golden Bauhinia Square), 침사추이 워터프런트(Tsim Sha Tsui's Water Front)를 따라 형성된 도시공간은 많은 관광객이 방문하는 아름다운 밤의 도시로 변모했다.

Hong Kong, the City Lights Never Goes Out

Hong Kong's nights are a symphony of light, and the city has earned a reputation as a colorful city of lights. Before the implementation of the Harbor Lighting Plan, which focused on Victoria Harbor, the city was affected by light pollution, as competing light displays from different buildings created a dazzling yet chaotic display.

However, after the implementation of the light plan, the urban spaces along the Avenue of Stars, Hong Kong Cultural Center, Wan Chai, Golden Bauhinia Square, and Tsim Sha Tsui's Water Front have been transformed into a beautiful nighttime city that is visited by many

밝지만 더 밝아야 할 곳은 더 밝게 하고, 어두워야 하는 곳은 빛을 배경 삼아 사람들이 즐길 수 있는 프로그램을 덧입혀서 도시의 빛 공해 해결은 물론 모든 사람이 즐겨 찾는 관광명소가 된 사례이다.

홍콩의 '빛의 심포니(Symphony of Light)'는 저녁 8시부터 10여 분 동안 홍콩의 밤을 화려하게 빛으로 물들인다. 이 빛의 심포니는 전 세계에 '환상의 도시 홍콩'의 명성을 알리는 새로운 도시브랜드를 창출했다.

tourists. By brightening up areas that need to be brighter and darkening areas that need to be darker, and adding programs that people can enjoy against the backdrop of the light, the city has become a tourist attraction for everyone, not to mention a solution to light pollution.

Hong Kong's 'Symphony of Light' lights up the city's nights for approximately 10 minutes starting at 8 p.m. to 10 p.m. The Symphony of Light has created a new city brand that has spread the reputation of the 'City of Illusions' around the world.

밤의 활기를 창조하는 홍콩 빛의 심포니 | Hong Kong's Symphony of Light creates a night vibrancy

빛이 절제된 품격의 도시, 파리

파리(Paris)의 밤에는 백색이나 청색이 없다. 낮에는 에펠탑(Eiffel Tower)을 랜드마크로 특정 대상에 집중된 경관을 보여주고, 나머지 지역은 오스만 양식의 통일된 층고와 파사드형 건축물로 조화로운 도시 이미지를 보여준다. 낮에 인지되는 이미지는 지역별 특성이나 역사성이 잘 발견되지 않는다고 볼 수 있다.

그러나 파리의 밤 풍경은 다르다. 파리를 밝히는 빛의 색감은 따뜻하다. 런던(London)은 따뜻함과 다르게 조금은 차가운 빛의 색(3,000K~5,000K)을 사용하는데, 파리는 도시의 역사성을 드러내기 위해서 따뜻한 느낌의 빛(1,850K~2,000K)을 통일되게 적용하고 있다.

Paris, The City of Dignity with Understated Light

There is no white or blue in Paris at night. During the day, the view is focused on the Eiffel Tower as a landmark, while the rest of the city is a harmonious image of Haussmannian architecture with uniform floor levels and facades. The perceived image during the day is one that does not reveal much in the way of local character or history.

Paris at night, however, is a different story. The color of the light that illuminates Paris is warm. Unlike London, which uses a slightly cooler light color (3,000K to 5,000K), Paris uses a uniformly warm light color (1,850K to 2,000K) to showcase the city's history.

따뜻한 빛으로 물들어 가는 프랑스 파리의 야경
파리시 랜드마크 에펠타워

Nightscape bathed with warm light, Paris, France
Paris landmark Eiffel Tower

석양 무렵의 노을빛인 나트륨램프*에 의해서 회색빛의 도시가 밤에는 오랜 역사적 이미지의 황금빛 도시로 변모한다. 만약 다양한 컬러(Color)가 존재했다면 도시의 역사성을 부각시키기보다는 혼잡한 도시 파리의 이미지로 훼손됐을 것이다. 단순한 컬러, 조화로운 빛 연출이 파리의 도시적 가치를 새롭게 창출해 낸 것이다.

At sunset, sodium lamps* transform the gray city into a golden city at night, a historical image. If there were a variety of colors, it would have undermined the image of Paris as a bustling city rather than highlighting the city's history. The simple colors and the harmonious play of light create a new urban value for Paris.

*나트륨램프: 원통형 유리관 속에 희유기체(稀有氣體)와 나트륨증기를 넣은 등. 방전하면 강한 노란색 빛을 발산하는 램프

*Sodium lamp: A cylindrical glass tube containing noble gases and sodium vapor. When discharged, the lamp emits a strong yellow light.

프랑스 리옹시 시청사 빛의 아트 | Light Art of Lyon City Hall, France

따뜻한 마음을 전하는 도시, 리옹

세계적인 빛의 도시 리옹

리옹(Lyon)시를 이야기할 때 먼저 제시되는 타이틀이 있다. 바로 '세계 빛의 도시 리옹'이다. 리옹시는 도시에 빛을 연출하여 새로운 모습을 만들어 간 대표적인 도시이다. 리옹은 파리에서 약 500km 떨어져 있는 도시로, 이탈리아나 스위스 등 유럽으로 가는 길목에 있다. 따라서 전시, 컨벤션, 물류의 유통단지로 유명하다. 많은 사람이 리옹시를 찾지만 저녁이 되면 대부분 화려한 도시 파리로 떠나간다.

관광화를 위해 기획된 빛의 도시

리옹시는 관광사업을 활성화할 방안으로 리옹시에 관광객이 머물게 할 방안을 찾았다. 1989년에 미셸 느와르(Michel Noir) 시장이 "리옹시에 조명을 입혀 사람들이 밤의 시간에도 머물도록 하겠다."라는 선거 공약을 내세워 시장에 당선됐다.

이 전략을 기본으로 도시의 조명이 처음으로 기획됐다. 도시의 기본 축인 론강(Rhône River)과 손강(Saône River)의 중요한 수변 축을 빛으로 연출하고, 녹지와 강을 어둠으로 보존하면서 두 개의 강변과 포비에르 바실리카 성당(La basilique Notre Dame de Fourvière)을 중심으로 도시 야경을 특징 있게 기획했다.

Lyon, the City that Gives Warm Heart

Lyon, the Global City of Lights

When talking about the city of Lyon, there's a title that comes up first. It is Lyon, the City of Lights. The city of Lyon is a prime example of a city that has transformed itself by using light to create a new look. Lyon is located about 500 kilometers from Paris and is on the way to the rest of Europe, including Italy and Switzerland. As such, it is known for its exhibition, convention, and distribution centers. Many people visit the city of Lyon, but most of them leave in the evening for the glittering city of Paris.

The City of Lights Designed for Tourism

The city of Lyon was looking for a way to boost tourism and keep tourists in the city. In 1989, Mayor Michel Noir was elected mayor with a campaign promise: "I will light up the city of Lyon so that people will stay after dark."

The city's lighting was first planned based on this strategy. The city's main waterfront axes, the Rhône and Saône rivers, were illuminated, while preserving the green spaces and rivers in darkness, and characterizing the city's nighttime landscape around the two rivers and the basilica of La basilique Notre Dame de Fourvière.

리옹 빛 축제 | Lyon Festival of Lights (La Fête des Lumière)

시드니 오페라하우스 | Sydney Opera House

감성의 빛을 담아내는 도시, 시드니

비비드 시드니(Vivid Sydney)

매년 5월, 호주 시드니(Sydney) 항구와 주변 지역은 찬란한 빛의 설치물과 미디어파사드를 통해 항구의 주요 공간이 매력 있는 빛의 구조물로 형상화된다. 오페라하우스의 건축구조물은 또 다른 빛의 캔버스가 되어 전 세계에서 한 해에 230만 명 이상의 관광객을 유치하는 도시가 됐다.

     도시를 온통 빛으로 물들여 새로운 도시가 탄생한다. 역동적인 빛으로 사람들을 춤추게 한다. 이곳에서 사람들은 고단한 일상을 내려놓고 눈앞에 펼쳐지는 환상적인 빛을 즐긴다. 비비드 시드니(Vivid Sydney)를 통해 딱딱한 도시공간이 부드럽고 편안한 관광지가 될 수 있다는 사례를 볼 수 있다.

     나는 조명디자이너로 초청받아 시드니를 방문했다. 시드니에서 빛 하나하나의 모습을 관찰할 수 있었고, 시드니 오페라하우스 건너편 항구에서 비추어지는 프로젝션 매핑(Projection Mapping) 비비드 시드니 미디어파사드 쇼를 세세하게 볼 수 있었다.

Sydney, the City that Captures Light of Emotions

Vivid Sydney

Every May, the harbor and surrounding area of Sydney, Australia, is transformed into a mesmerizing light installation and media façade that transforms the harbor's main space into a mesmerizing light structure. The architectural structure of the Opera House becomes another canvas of light, attracting more than 2.3 million tourists a year from around the world.

     When the whole city is bathed in light, a new city is born. Dynamic light makes people dance. Here, people put aside their daily routine and enjoy the spectacular light show in front of them. Vivid Sydney is an example of how a hard urban space can become a soft and comfortable tourist destination.

     I was invited to visit Sydney as a lighting designer, and I was able to observe every single light in the city and see the details of the Projection Mapping Vivid Sydney media facade show on the harbor across from the Sydney Opera House.

비비드 시드니 | Vivid Sydney

활기 넘치는 도시를 창출하는 빛의 예술

홍콩이 무분별한 빛 환경을 개선했고, 파리시는 도시의 역사성을 드러내기 위해 밤 빛의 색감을 고려한 빛 연출에 관심을 가졌다. 시드니와 리옹시는 새로운 빛 연출 기획을 통해 도시를 방문한 관광객과 시민들이 더 많은 시간을 도시에 머물게 하여 도시경제 활성화를 이루어 낸 성공적인 사례로 볼 수 있다. 빛은 여러 가지 목적에 따라 다양하게 활용될 수 있다.

　　　　밝아야 할 곳은 밝히고 어두워야 할 곳은 어두워야 한다는 빛 연출의 기본 규칙을 통해 안전하고 쾌적한 도시의 밤을 만들고, 일정 시간에 연출되는 미디어파사드 쇼는 활기찬 도시를 만들어 낸다.

The Art of Light to Create Vibrant Cities

Hong Kong has enhanced its expansive lighting infrastructure, while Paris is interested in implementing color-conscious lighting at night to highlight the city's history. Sydney and Lyon serve as successful examples of how new lighting design initiatives have promoted longer stays by tourists and residents, thereby stimulating the local economy.

　　　　Light can be utilized for many different purposes. The basic rules of light design - bright where it should be bright and dark where it should be dark - create safe and comfortable city nights, while timed media facade create vibrant cities.

시드니 하버 브리지 | Sydney Harbor Bridge

나에게 도시는 꿈을 펼쳐내는 무대이다.
나는 도시라는 무대 위에 주연과 조연을 설정하고
나는 빛으로 이 도시만의 이야기를 들려준다.

Cities serve as the stage to unfold my dream,
where I carefully position the main and supporting characters
I harness light to narrate the city's unique story.

빛 공해를 개선한 서울시 야간경관 | Improving nightscape light pollution, Seoul, 2024

Chapter III
빛을 담다: 光

CAPTURING THE LIGHT: LIGHT

매력적인 빛, 도시의 가치를 찾다

밤의 도시경관을 단순히 조명을 밝히는 일이라고 생각하는 사람이 많다. 이제는 빛으로 도시 공간의 감성을 찾아야 한다는 사고의 전환이 필요한 때다. 도시에는 역사와 문화의 가치를 간직한 다양한 장소가 있다. 이런 도시 공간이 산업화와 발전이라는 이름으로 빛의 양적인 팽창을 거듭해 고유한 밤의 경관을 잃어가고 있다.

Attractive light, finding value in the city

Many people think of urbanism at night as simply lighting up the cityscape. It is time to rethink the way we use light to find the emotion of urban spaces. There are various places in cities that hold the value of history and culture. In the name of industrialization and development, these urban spaces have been subjected to a quantitative expansion of light, and are losing their unique nighttime scenery.

제3장에서는 빛을 연출하고 디자인한 현장의 작품을 담았다. 우리는 현장에서 빛을 디자인하기에 앞서 현장 테스트를 통해서 빛의 가치를 재해석해 내려고 노력했다.

모든 빛 연출은 현장 중심으로 계획되었고, 문제도 현장에 있고 그에 대한 해답도 현장에 있다고 보았다. 그래서 우리는 무대의 작품 완성도를 높이기 위해 빛 디자이너들과 현장에서 끊임없이 리허설하면서 중지를 모았다. 이런 과정에서 잘못된 계획도 바로잡히고, 때로는 기발한 발상이 창출되기도 했다.

In Chapter 3, we showcase our work in the field where we created and designed lighting. Our approach involved field testing to redefine the value of light before designing it on site.

Each lighting production was tailored to the specific site, as we believed that the problems were there and the solutions were there. We continuously collaborated with lighting designers in the field to refine our work. This process enabled us to correct any flawed plans and generate new, brilliant ideas along the way.

한국은행 화폐금융박물관 빛의 디자인, 서울시 건축상 수상 | Light Design, Bank of Korea Museum of Money and Finance, Seoul Architecture Award, 2005

밤을 비우다:
아름다운 한강의 밤 '한강 빛의 르네상스'

EMPTY THE NIGHT:
LIGHT MASTER PLAN OF 'HANGANG RIVER RENAISSANCE'

한강르네상스 빛의 마스터플랜
LIGHT MASTER PLAN OF HANGANG RIVER RENAISSANCE

서울시 여의도한강공원 빛 연출
LIGHT PRODUCTION OF YEOUIDO HANGANG PARK
IN SEOUL, KOREA

서울시 난지한강공원 빛 연출
LIGHT PRODUCTION OF NANJI HANGANG PARK
IN SEOUL, KOREA

낮보다 아름다운 서울과 한강의 밤
빛을 만들기 전에
어두운 서울과 한강을 먼저 생각했다.

Seoul and the Hangang River at night is more beautiful than during the day.
Before creating light,
We first thought about the dark Seoul and Hangang River.

한강르네상스 빛의 마스터플랜

한국은 2002년 월드컵이라는 세계적 행사를 개최한 이후 밤에 대한 인식이 달라지면서 야경을 도시 이미지 개선에 중요한 요소로 보려는 발상 전환이 이뤄졌다. 서울시도 국제도시의 위상에 걸맞게 한강을 대상으로 낮과 밤의 경관을 새롭게 개선하겠다는 '서울 한강르네상스' 사업을 시행하게 됐다.

종합적인 '한강르네상스 빛의 기본계획' 수립

기존에 설치된 한강교량과 한강공원의 밤조명은 전체적인 계획이 없는 상태에서 각기 다른 조명 기준으로 조성됐다. 일부는 과도하게 설치된 조명과 상향으로 설치된 조명 때문에 서울 한강 주변 밤의 풍경이 오히려 빛 공해로 훼손된 상태였다. 이는 그동안 야간경관 사업이 밤을 밝히 밝히는 기능에만 집중됐기 때문이다. 단기적인 사업 중심의 일회성으로 볼거리를 제공하기 위한 테마파크 조성으로 전락해 버린 결과였다. 다시 말하면 대부분 야간 경관계획이 단기적인 안목으로 진행되어 밤을 배려하는 종합적인 계획이 전혀 없었다는 것이다.

한강대교 23개와 11개의 한강공원을 종합한 빛 연출

우리는 '한강르네상스 계획*'에서, 서울시에서 관할하는 한강 23개 교량과 11개 한강공원, 한강 주변 500m 거리에 위치하는 민간 건축물에 대해서 하나의 콘셉트(Concept)로, 5개 권역별로 빛의 기본계획을 만들어서 종합적인 '한강르네상스 빛의 기본계획'을 만들어 냈다.

우리는 서울의 역사, 문화, 자연의 모습을 담고 있는 한강에 대해서 종합적인 기획으로 접근했다. 한강 본연의 모습을 지키되, 보여주기보다는 낮추거나 숨기고, 드러내기보다는 느낄 수 있는 밤의 정취를 담겠다는 감성적인 자세로 접근했다. 이를 위해 서울의 한강과 주변 건축물, 한강공원과 한강의 교량 본연의 구조미를 강물 속에 비추어지거나 반사되는 빛의 모습에서 찾아내고자 했다.

LIGHT MASTER PLAN OF HANGANG RIVER RENAISSANCE

Since South Korea hosted the 2002 World Cup, the country's perception of night has changed and the idea of nighttime has become an important factor in improving the city's image. In line with its status as an international city, Seoul launched the Hangang Renaissance Project, which aims to improve the daytime and nighttime scenery along the Hangang River.

Establishment of a Comprehensive 'Light Master Plan of Hangang River Renaissance'

The nighttime lighting of the existing Hangang River bridges and Hangang Park was created with different lighting standards without an overall plan. Some of the lights were installed excessively and others were installed upward, resulting in light pollution that spoiled the nighttime scenery around the Hangang River in Seoul. This is because the night view project has been focused only on the function of brightening up the night. This resulted in the creation of a theme park to provide a short-term, business-oriented, one-time attraction. In other words, most of the night view plans were carried out with a short-term perspective, and there was no comprehensive plan to consider the night.

Comprehensive Light Production for 23 Hangang Bridges and 11 Hangang Parks

In Hangang Renaissance Plan*, we created a comprehensive 'Light Master Plan of Hangang Renaissance' by creating a master plan for each of the 23 bridges, 11 Hangang Parks, and private buildings located within 500 meters of the Hangang River under the jurisdiction of the Seoul Metropolitan Government as one concept for each of the five areas.

We approached the Hangang River, which contains the history, culture, and nature of Seoul, as a comprehensive plan. We approached it with an emotional attitude to preserve the original appearance of the Hangang River, but to lower or hide it rather than show it, and to capture the nighttime atmosphere that can be felt rather than revealed. To this end, we tried to find the structural beauty of the Hangang River in Seoul, the surrounding architecture, Hangang Park, and the bridges of the Hangang River through the light that shines or reflects in the river.

기존의 야간 경관계획이 단지 교량과 건축물을 보여주기 위한 개별적인 '빛 중심의 계획'이었다면, 우리의 계획은 대한민국 수도 서울의 중심축인 한강의 밤 풍광을 종합적으로 드러내는 기획이었다. 즉, 한강의 밤의 어둠을 지켜내어 '한강의 넓고 긴 선형'을 빛으로 드러낸다는, '아름다운 한강의 밤 풍광'을 위한 종합적인 계획이었다.

성장 동력의 서울과 한강 이미지 부각

특히 한강만의 장소성을 표현하기 위해서 빛이 아닌 밤을 담기 위한 개념으로, 빛을 보태어 비추는 것이 아니라 빛을 낮추고 어둠을 지켜주려는 '빛의 절제 개념'을 도입했다. 이 같은 빛의 절제 개념은 한강 본연의 생태적인 기능을 회복시키겠다는 '한강 밤의 모습 보존계획'과 서울의 성장 동력으로서 한강의 기능을 창출하겠다는 '한강 빛 환경 형성계획' 두 사업 계획을 결합한 기획이었다.

Whereas previous night view plans were individual 'light-centered plans' to showcase bridges and architecture, our plan was a comprehensive plan to reveal the night view of the Hangang River, the centerpiece of Seoul, the capital of South Korea. In other words, it was a comprehensive plan for a 'beautiful Hangang River night view' that preserves the darkness of the Hangang River at night and reveals the 'wide and long linearity of the Hangang River' in light.

Emphasize Seoul and the Hangang River as Growth Engines

In particular, in order to express the sense of place of the Hangang River, the concept of 'light moderation' was introduced to capture the night, not to shine brighter, but reduce the light and protect the darkness. This concept of light restraint was a combination of two project plans: the 'Hangang River Night Preservation Plan' to restore the ecological function of the Hangang River and the 'Hangang River Light Environment Formation Plan' to create the function of the Hangang River as a growth engine for Seoul.

서울시 잠두봉에서 바라본 한강 | Hangang River viewed from Jamdubong, Seoul

*서울시 한강르네상스계획: 치수(治水)를 바탕으로 적극적 이수(利水)를 통해 한강을 맑고 매력이 넘치는 서울의 상징으로 조성시키겠다는 2006년 서울시에서 계획한 한강 종합정비계획

*Hangang Renaissance Plan: A comprehensive Hangang Renaissance plan planned by the Seoul Metropolitan Government in 2006 to make the Hangang River a clear and attractive symbol of Seoul through controlling water based on the principle of utilizing water.

대한민국 심장으로서의 한강 빛 이미지 연출

서울 한강의 평균 강폭은 약 1,000m, 서울시 경계는 약 41.5km의 길이, 서울 전체 면적의 약 7%로써, 우리는 큰 강을 보물로 지녔다. 영국 템즈강, 프랑스 센강, 리옹의 론강, 헝가리 다뉴브강은 한강에 비해 폭이나 길이에 훨씬 못 미친다. 한강은 대한민국 수도 서울의 상징으로, 다양한 이미지를 담고 있다. 우리는 처음에 어떻게 빛의 계획을 세울 것인가? 라는 무게감을 안고 출발했다. 한강을 이해하고 느껴가면서 새로운 관점으로 한강의 빛을 새롭게 디자인해 나갔다.

Create a Light Image of the Hangang River as the Heart of Korea

With an average river width of about 1,000 meters and a length of 41.5 kilometers bordering Seoul, or about 7% of the city's total area, we are blessed with a great river. The Thames in England, the Seine in France, the Rhone in Lyon, and the Danube in Hungary are nowhere near the width or length of the Hangang River. The Hangang River is a symbol of Seoul, the capital of South Korea, and is associated with many different images. We started out with the weight of how to plan the light? We started out with the weighty question of how to plan the light. As we understood and felt the Hangang River, we redesigned the light from a new perspective.

서울 63빌딩에서 조망되는 한강 | View of the Hangang River from 63 Building in Seoul

한강르네상스 빛의 디자인 방향

우리는 서울의 역사, 문화, 자연을 지켜주고자 하는 마음으로 한강 본연의 밤과 낮의 모습을 담아내고자 했다. 그래서 빛을 새롭게 만들기보다는 빛을 비우고자 했다. 기존의 빛을 지우고 낮추니 서울 한복판을 관통하는 한강의 윤곽과 함께 한강의 위엄이 새롭게 나타났다.

Light Design Directions of Hangang Renaissance

We wanted to capture the essence of Hangang River at night and during the day to preserve the history, culture, and nature of Seoul. We wanted to empty the light rather than create new light. By erasing and lowering the existing light, the majesty of the Hangang River emerged, along with the contours of the river that runs through the center of Seoul.

빛의 물결 위에 피어 오르는 서울의 밤 | Night of Seoul rising above the waves of light

한강의 자연 요소인 청야(淸夜)

서울을 관통하는 크고 넓은 한강다운 모습, 이런 한강의 정체성을 빛으로 연출하고자 한강 밤의 콘셉트로 청야(淸夜)의 개념을 적용했다. 청야는 한강의 자연적 요소인 물과 바람, 인문적 요소인 한민족의 장구한 역사를 안고 흐르는 한강의 도도한 기상이다. 이를 드러내기 위해 거대한 한강의 물줄기는 어둠으로 서울의 중심축을 만들어서 한강 본연의 밤 이미지 회복을 연출하고 동시에 23개 한강교량은 한강의 물결 위에 보이게 연출하여 서울의 역동성을 최소한의 빛으로 드러나게 연출했다.

한강르네상스 빛의 개념인 '청야(淸夜)'를 통해 서울의 동서축을 형성하는 한강 본연의 구조인 W 자(字) 형태의 선형을 밤의 어둠으로 연출할 수 있는 빛의 종합적인 가이드라인을 먼저 계획했다.

Chungya, a Natural Element of the Hangang River

The concept of the Hangang River's night was created to represent the identity of the Hangang River, a large river that runs through Seoul. Cheongya symbolizes the natural elements of the Hangang River, including water and wind, as well as the historical connection of the Korean people. To reveal this, the massive water stream of the Hangang River was darkened to serve as the central axis of Seoul, restoring the original nighttime appearance of the river. Additionally, the 23 Hangang River bridges were illuminated to appear as if they are floating above the waves, highlighting the energy of Seoul with minimal lighting.

Through the Hangang River Renaissance concept of light, 'Chungya', we first planned a comprehensive lighting guideline to create a W-shaped linear line, the original structure of the Hangang River, which forms the east-west axis of Seoul, in darkness at night.

서울의 중심을 가로지르는 거대한 물줄기, 한강 | Massive waterway that runs through the heart of Seoul, Hangang River

개선 전: 붉게 확산되는 혼탁한 조명의 한강 | Before improvement: Murky night of Hangang River hidden by the spreading red light

개선 후: 맑고 깨끗한 빛으로 물든 밤의 한강 | After improvement: The Hangang River at night, dyed in clear and pure light.

서울 한강교량 빛의 디자인

우리는 기존에 설치되어 있는 교량의 조명 콘셉트의 기조를 유지했다. 대신 일부 누수되는 빛을 개선하는 동시에 과도한 조명은 전체를 철거하고, 새롭게 설치하는 빛은 주변 지역에 맞추어서 빛을 디자인했다. 한강 교량을 대상으로 한강의 다리가 지닌 각각의 구조미가 일몰에서 밤 11시까지만 한강의 물결 위에서 빛이 연출되도록 빛을 디자인했다.

Light Design for Hangang River Bridges in Seoul, Korea

Leaky lights and unnecessary lights were removed, and new lights were carefully installed to harmonize with the surroundings. For the bridges on the Hangang River, the new lighting scheme was specifically tailored to accentuate the structural beauty of each bridge, casting a delightful reflection on the river from sunset to 11 pm.

한강 성산대교 빛의 디자인 | Light Design of Seongsan Bridge on the Hangang River

서울 23개 한강교량 빛의 디자인 가이드라인

한강교량의 빛이 한강으로 유입되지 않도록 하고, 한강 수면이 하나의 큰 거울이 되어 한강교량의 위용을 빛으로 담아내게 했다. 이를 위해 한강교량의 구조미와 상징성을 나타내기 위한 빛의 디자인 전략이 먼저 수립되었고, 기존에 과도하게 설치되어 있던 교량 조명에 대해서도 조명시설을 과감하게 철거하거나 빛을 낮출 수 있도록 조명을 재조정했다.

    이 과정에서, 빛 차단막을 설치하여 새는 빛(Spill Light)을 차단하고, 조명 컬러는 교량 본연의 마감 색상이 돋보일 수 있도록 만들었다. 교량에 사용하는 컬러의 변화 속도는 빠른 변

Guidelines of the Light Design of 23 Hangang River Bridges in Seoul

The Hangang River Bridge now stands as a majestic symbol, with its light carefully designed to showcase its beauty and significance. By controlling the flow of light, the bridge's reflection on the water's surface creates a stunning spectacle. Through strategic lighting rearrangement, the excessive brightness has been reduced, preserving the bridge's allure.

    During this process, light shields were installed to block Spill Light, and the color of the lighting was chosen to

1 행주대교 | Haengjudaegyo Bridge

7 서강대교 | Seogangdaegyo Bridge

9 원효대교 | Wonhyodaegyo Bridge

2 방화대교 | Banghwadaegyo Bridge

3 가양대교 | Gayangdaegyo Bridge

4 성산대교 | Seongsandaegyo Bridge

5 양화대교 | Yanghwadaegyo Bridge

서울 한강 23개 대교 | 23 Major Bridges of the Han River in Seoul

6 당산철교 | Dangsandaegyo Bridge

8 마포대교 | Mapodaegyo Bridge

10 한강철교 | Hangang Railway Bridge

화보다 느리게, 보색대비와 원색의 색상 사용을 자제하여 한강교량의 위용이나 상징성이 훼손되지 않도록 했다.

complement the original finish color of the bridge. The pace of color change on the bridge was slower than fast, and the use of complementary contrasting and primary colors was avoided to ensure that the majesty and symbolism of the Hangang River Bridge was not compromised.

11 한강대교 | Hangangdaegyo Bridge

14 한남대교 | Hannamdaegyo Bridge

17 영동대교 | Yeongdongdaegyo Bridge

12 동작대교 | Dongjakdaegyo Bridge

15 동호대교 | Donghodaegyo Bridge

18 청담대교 | Cheongdamdaegyo Bridge

19 잠실대교 | Jamsildaegyo Bridge

20 잠실철교 | Jamsil Railway Bridge

21 올림픽대교 | Olympic Bridge

22 천호대교 | Cheonhodaegyo Bridge

13 반포대교 | Banpodaegyo Bridge

16 성수대교 | Seongsudaegyo Bridge

23 광진교 | Gwangjin Bridge

한강 교량 중에서 개선이 시급한 가양대교, 성산대교, 양화대교, 동작대교와 원효대교, 한강대교 등 6개 한강교량에 대한 빛 디자인 설계가 우선으로 시작됐다. 우리의 빛 개선 방향은 첫째, 기존에 조성된 한강교량의 조명 콘셉트를 유지하면서 일부에 발생하는 빛 공해를 개선해 나가는 방향으로 빛을 디자인했다. 둘째, 일부 과도한 기존의 한강교량 조명에 대해서는 전체를 철거하고 빛을 새롭게 디자인하는 방향으로 진행했다. 이런 빛 디자인 개선 작업을 통해 한강교량의 위용을 드러냈다.

We prioritized on improving the lighting design for six Hangang River Bridges: Gayang Bridge, Seongsan Bridge, Yanghwa Bridge, Dongjak Bridge, Wonhyo Bridge, and Hangang Bridge. Our main goal was to maintain the existing lighting concept while reducing light pollution. We also removed excessive lighting from some bridges and redesigned them. These improvements helped to showcase the beauty of the Hangang River Bridges.

가양대교(加陽大橋)

가양대교 상판의 하부 면을 업라이트(Up Light) 조명하는 MH1.8KW/h 투광기에서 발생되는 눈부심과 교량 측면 MH250W/h 투광기에 의해 교량을 얼룩지게 만드는 부분을 개선하여 가양대교 본연의 컬러감과 구조미를 연출

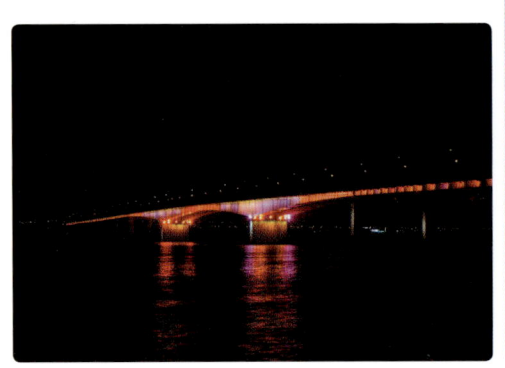

가양대교 빛의 디자인 | Light Design of Gayangdaegyo

성산대교(城山大橋)

상향광에 의한 빛 공해를 발생시키는 기존 성산대교 상판 측면 1.8KW/h 조명을 철거하고 교량 트러스 구조미를 부각시킬 수 있는 내부 투광 조명(MH250W/h, NH250W/h, MH400W/h, NH400W/h)각도 조절과 함께 교대 측면을 추가로 조명 보완하여 자연스럽게 성산대교 전체 구조미를 실루엣으로 연출

성산대교 빛의 디자인 | Light Design of Seongsandaegyo

양화대교(楊花大橋)

외부에서 조망 시 노출된 광원에 의해 눈부심을 발생시키고 있는 교량 난간 라인조명(Cold-Cathode Fluorescent Lamp)을 LED조명으로 교체 보완하고 교각 측면 투광 조명을 추가하여 교량 구조미를 연출. 교량 내부구조에 설치된 과도한 NH150W/h 조명은 철거하고 교각 외측부에 3200K(온백색) 조명을 간접 조명으로 설치하여 한강의 강물에 교량의 구조가 반사되는 효과를 연출

양화대교 빛의 디자인 | Light Design of Yanghwadaegyo

동작대교(銅雀大橋)

나트륨램프의 붉은색 도로조명과 교량 아치 구조를 나타내는 파란색 경관조명으로 아치 구조물에 얼룩짐 현상이 발생됨. 서로 다른 빛의 혼합적인 이미지를 개선하기 위해서 도로조명 가로등은 백색의 광원으로 가로등 조명기구만을 교체하고, 기존에 설치되어 있던 72개 조명의 파란색 필터(Blue Filter)를 제거하여 자연스럽게 교량 아치 구조 본연의 색감이 드러나도록 연출

동작대교 빛의 디자인 | Light Design of Dongjakdaegyo

원효대교(元曉大橋)

강변북로에서 조망 시 눈부심을 발생시키고 있는 MH250W/h, MH400W/h 투광 조명과 교량을 보색대비로 얼룩지게 만들어 내는 기존 조명시설에 대해서 전체를 새롭게 조성. 원효대교 교각과 아치형 구조가 본연의 미가 최소한의 빛으로 한강 수면에 담길 수 있는 빛을 디자인

원효대교 빛의 디자인 | Light Design of Wonhyodaegyo

한강대교(漢江大橋)

회색 빛 교량 트러스 구조에 차가운 파란색 컬러(Blue Color)를 적용하여 더욱 창백하고 음습한 경관을 조성시키고 있는 부분을 LED 간접조명(3000K)의 빛으로 한강대교 자체 구조미와 역사성을 연출. 빛은 운전자에게 눈부심이 느껴지지 않도록 교량 구조물 내부에서 업라이트(Up Light)로 연출하고 교량 측면 보행 공간을 다운라이트(Down Light)로 연출하여 어두웠던 보행 공간이 밝고 따뜻한 분위기공간으로 새롭게 조성

한강대교 빛의 디자인 | Light Design of Hangangdaegyo

서울시 11개 한강공원 41.5km의 빛 디자인

한강은 총연장 497.5km 중에서 서울특별시 행정구역(강동구 하일동~강서구 개화동)의 연장 41.5㎞, 면적은 39.9㎢, 8개 관할구청에 걸쳐서, 총 11개 한강공원이 한강변에 형성되어 있다. 한강공원은 시민과 관광객이 즐겨 찾는 서울의 대표적인 명소로 다양한 체험을 할 수 있는 수변 공간이다. 빛 연출은 한강의 회복이라는 주제를 고려하여 한강 주변 녹지 대부분의 밤 모습을 보존할 수 있도록 기획했다. 한강의 밤은 한강공원에서 지켜져야 할 중요한 부분이라고 생각했다. 공원에서는 빛 공해를 제거하고 좋은 빛으로 쾌적함을 높이기 위해서 최대한 공원 이용자의 시각에서 다른 빛이 유입되지 않도록 계획했다.

한강공원의 성격상 산책로와 자전거도로, 조경시설과 운동 공간으로 구분되는 입체적 공간 특성을 고려해서 빛을 디자인했다. 11개 한강공원은 진·출입부, 수변부 산책로, 운동시설, 자전거도로, 주차장 등 이용 특성을 반영한 빛 연출 가이드라인을 새롭게 수립하여 적용했다.

한강공원 공원 자연 그대로의 색을 담다

낮에는 녹색의 한강공원이 밤이면 공원조명(NH400W, 1950K)으로 인해 모두 붉은 갈색빛으로 물들어 있다. 우리는 한강공원의 푸른 수목이 밤이 되면 붉은 갈색으로만 조망되는 한강공원의 빛 공해 문제를 해결하고 싶었다. 하늘로 날아올라 분산되는 빛을 차단하여 에너지를 절감시키고, 하늘이 맑은 밤이면 별빛을 볼 수 있게 빛을 디자인하고, 빛의 색감 개선을 통해서 나무가 본연의 푸른 자연 색감을 나타낼 수 있도록 개선했다. 특별한 시설에 의해서가 아니라 빛의 원칙과 소신을 담겠다는 의지만으로 가능했다.

우리는 한강공원 조명개선을 위해 서울시 관계자와 현장에서 함께 문제점을 찾고 해결할 방안을 찾았다. 문제의 현장에서 빛에 대한 테스트를 진행하면서 9m 높이의 공원 조명기구 위로 올라갔다. 조명기구를 위에서 교체하고 조명기구를 산책로 방향으로 다시 조정했다. 강 건너편에서 한강의 산책로를 바라보았다. 우리가 교체한 따뜻한 백색의 빛이 주변의 갈색의 빛 사이로 나무 본연의 녹색 빛으로 빛났다. 빛은 문제의 현장에서 함께 고민할 때 문제를 개선할 수 있다는 사실을 한강 빛의 테스트를 통해 체험했다.

Light Design for 41.5 kilometers of Seoul's 11 Hangang Parks

Of the total length of 497.5 kilometers, 41.5 kilometers are in the administrative district of Seoul (Haildong, Gangdong-gu to Gaewha-dong, Gangseo-gu), and 39.9 square kilometers are in eight jurisdictions, and a total of 11 Hangang Parks are located along the Hangang River. Hangang Park is a waterfront space where citizens and tourists can enjoy various experiences as a representative attraction of Seoul. The light production was designed to preserve the nighttime appearance of most of the greenery along the Hangang River, in keeping with the theme of restoring the river. We felt that the nighttime of the Hangang River was an important part of Hangang Park that should be preserved. In order to eliminate light pollution and enhance comfort with good light, the park planned to keep other lights out of the view of park users as much as possible.

The light design takes into account the three-dimensional nature of the parks, which are divided into walking and biking paths, landscaping, and exercise areas. The 11 Hangang Parks have newly established and applied light production guidelines that reflect the characteristics of their use, such as entrances and exits, waterfront trails, exercise facilities, bicycle paths, and parking lots.

Hangang Park Capturing the Colors of the Park's Natural Environment

Hangang Park is green during the day, but at night, the park lights (NH400W, 1950K) turn it all reddish brown. We wanted to solve the problem of light pollution in Hangang Park, where the green trees are only visible as reddish brown at night, by blocking the light that flies up to the sky to save energy, designing the light so that starlight can be seen on clear nights, and improving the color of the light to show the natural blue color of the trees. This was not possible with any special facilities, but only with the will to incorporate the principles of light and our own opinions.

To improve the lighting of Hangang Park, we worked on-site with Seoul officials to find problems and solutions. While testing the light at the site in question, we climbed 9 meters up the park's luminaires. We replaced the fixtures from above and reoriented them toward the promenade. From across the river, we looked at the Hangang River walkway. The warm white light we'd replaced shone through the brown light of the surroundings, illuminating the green of the trees. The Hangang River Light test showed us that light can improve problems when we work together at the site of the problem.

한강공원을 붉고 탁하게 만든 과도한 조명 | The excessive lighting that made the Han River Park red and murky

부드러운 빛으로 개선한 후, 나무의 녹색 빛이 보이는 한강공원 | The Hangang River Park after improvement with soft lighting, where the green light of the trees is visible

현장 테스트 후, 빛의 개선 사항을 서울시 관계자에게 설명 | After the on-site test, explaining the improvements of the lighting to the Seoul city officials

강 건너 조명 테스트가 진행된 곳에서 보이는 한강 자연의 빛 | The natural light of the Han River visible from the location where the lighting test was conducted across the river

하늘공원에서 조망되는 한강 | Hangang River seen from Sky Park

주간의 녹색이 모두 갈색으로 변한 한강공원의 밤 | Night of Hangang Park, where the green of the day has all turned brown

하늘과 강으로 향하는 빛공해를 개선한 푸른 빛의 한강공원 | Blue Hangang Park, improving light pollution toward the sky and river

잠실 한강공원 주간 경관 | Jamsil Hangang Park viewed during the day

잠실 한강공원 개선 전 야경 | Night view of Jamsil Hangang Park before improvement

잠실 한강공원 개선 후 야경 | Night view of Jamsil Hangang Park after improvement

한강 주변에 인접한 건축물 빛의 디자인

한강에 인접한 건축물에 대해서 나는 한강의 지형구조인 W 형태를 담아내는 빛의 그릇으로 밤의 모습을 상상했다. 한강 W 형태로 굴곡진 부분에 위치한 여의도와 강남, 강북지역의 건축물은 현대화된 모습을 지니고 있다. W 형태로 굴곡부에 위치하는 건축물 중에서 강변에 접한 건축물의 특징을 빛으로 디자인해 냈다. 빛의 가이드라인을 구축해서 한강의 역동적인 모습이 한강의 물빛으로 담아낼 수 있도록 빛의 스카이라인을 구상했다.

  빛을 담아내는 계획과 함께 서울의 밤 풍광을 실루엣으로 연출하여 서울의 도시경관을 인상적으로 표현하고자 한 공간이 있다. 바로 한강여의도공원 주변이었다. 여의도는 한강 주변에서 서울의 대표적인 자연경관인 남산을 가까이에서 온전하게 바라볼 수 있는 공간이다. 이곳에서는 한강 주변의 건축물 조명을 지양(止揚)하는 조명 가이드라인을 수립했다. 남산의 뒤쪽 도심에서 형성된 밤 조명에 의해서 남산의 윤곽이 자연스럽게 어둠 속에서 실루엣으로 보이도록 하여 서울이라는 도시의 중심 무대가 되도록 했다. 그리고 그 앞으로 띠를 두른 듯 강변도로의 불빛이 생동감 있게 형성되고, 도심 건물이 뿜어내는 찬란한 불빛으로 밤의 시각회랑(視覺回廊, Visual Corridor)이 자연스럽게 조성되도록 했다.

  한강 밤의 빛 디자인은 한강에 새로운 빛을 창출하는 일이다. 기왕의 빛을 제거하기도 하고, 새로 빛을 밝히기도 하지만 무엇보다도 조화의 원칙을 지켜가야 한다. 밤의 참모습을 훼손시키는 가장 손쉬운 방법은 밤을 낮처럼 환하게 밝히면 된다. 그러나 그 순간부터 다시는 본연의 밤을 만날 수 없다. 우리는 이런 밤의 빛 디자인 금칙(禁飭)을 지켜가면서 조심스럽게 밤의 빛 디자인을 해나갔다. 빛을 개선한 뒤 밤의 한강과 도시는 빛이 차분하게 어우러지고 위엄있는 밤의 모습을 드러냈다.

Architectural Light Design Adjacent to the Hangang River

For the buildings adjacent to the Hangang River, I envisioned them at night as vessels of light that reflect the W-shape of the river's topography. The buildings in Yeouido, Gangnam region, and Gangbuk region, located in the W-shaped bend of the Hangang River, have a modernized look. Among the buildings located in the W-shaped bend, we designed the features of the buildings facing the river with light. By establishing light guidelines, a light skyline was envisioned to capture the dynamism of the Hangang River with the color of the river's water.

  Along with the plan to capture light, there was one area where we wanted to create a silhouette of the night view of Seoul to impressively express the cityscape of Seoul. This was the area around Yeouido Park on the Hangang River. Yeouido is a space along the Hangang River where you can see Namsan, one of Seoul's most iconic natural landscapes, up close and personal. Here, lighting guidelines were established that prohibited the lighting of buildings along the Hangang River. The outline of Namsan Mountain is naturally silhouetted in the dark by the night lights from the city center behind Namsan Mountain, making it the center stage of the city of Seoul. In front of it, the riverside roads are vividly illuminated like a band, and the brilliant lights of the city center buildings create a visual corridor at night.

  The design of the Hangang River Night Light is about creating a new light on the Hangang River. It can be a matter of removing old light or adding new light, but the principle of harmony must be observed above all. The easiest way to destroy the night is to make it as bright as day. But from that moment on, the night will never be the same again. We carefully designed the night light while observing these rules of night light design. After improving the light, the Hangang River and the city at night are calmly illuminated, revealing the majesty of the night.

한강 주변 건축물 경관조명 조사 | Investigation of architectural lighting around the Hangang River

한강 이촌지구 민간건축물 경관조명 테스트 | Hangang River Ichon District private building architectural lighting test

조명 물리량 측정 | Measurement of lighting physical quantities

주간의 한강변 경관 | Daytime view of the Hangang River Landscape

야간의 한강변 경관 | Nighttime view of the Hangang River Landscape

야간의 한강변 조명 물리량 계측 | Measurement of light intensity of Hangang River at night

서울시 한강 밤의 디자인

밤을 새로운 개념으로 이해하고 빛 연출로 개선하려 할 때 고정된 관념과의 충돌은 불가피하다. 한 예가 한강공원 반포지구 공원조명 개선이었다. 반포한강공원이 너무 어둡다는 시민의 민원에 따라 공원등 추가설치가 필요했다.

우리는 이 작업에서 기존 공원조명(나트륨램프 400W/h, 1,950K의 전반 확산 조명)을 소비전력이 5배가 더 낮은 조명으로 교체(메탈램프 70W/h, 3,000K의 컷오프 조명)해야 한다는 새로운 빛 디자인을 제안했다. 그러자 서울시는 우리가 제안한 낮은 소비전력의 조명이 밤에 안전상의 문제가 될 수 있다고 이의를 제기했다. 빛은 단순히 소비전력이 높다고 밝은 조도를 확보할 수 있는 것이 아니다. 먼저 조명기구의 배광(Light Distribution) 특성과 연출 방법이 검토돼야 한다. 현재 한강의 밤을 가장 크게 훼손시키는 부분이 바로 안전 확보를 위해 공원에 설치한 400W/h 고출력 전반 확산 조명이었다. 위로는 하늘을 향해, 아래로는 강물로 빛을 쏟아붓는다. 이렇게 되자 주변 주거지역 아파트 실내 공간으로 빛이 유입되어 사람의 숙면을 방해하는 빛 공해에 시달리고 있었다.

우리는 상암동 월드컵 축구경기장의 1,000W/h 조명과도 같은 조명을 공원에 400W/h의 조명을 사용한다는 것 자체가 문제라고 판단했다. 우리는 서울시 관계자에게 실제적인 빛의 조명테스트를 통해서 문제점을 일깨워 주고 싶었다. 1차 조도 계산 시뮬레이션을 통해서 데이터로 조도 확보가 가능하다는 사실을 제시하고, 2차 한강공원 반포지구 서래섬 현장에서 이동용 무소음 발전기를 사용한 현장실험으로 우리의 빛 디자인을 검증했다. 빛의 현장실험 결과는 충분한 노면 조도의 확보와 함께 소요전력의 절감과 그동안 문제가 되었던 한강공원 조명의 과도한 빛에 의해 발생하던 눈부심과 상향된 빛을 제거하여 쾌적한 한강의 밤 모습을 연출할 수 있음을 현장실험으로 확인했다.

Design at Night on the Hangang River in Seoul

When trying to understand night in a new way and improve it with light production, it is inevitable to clash with fixed ideas. One example is the improvement of park lighting in the Banpo district of Hangang Park. Following complaints from citizens that Banpo Hangang Park was too dark, additional park lights were needed.

We proposed a new light design that called for replacing the existing park lights (sodium lamps 400W/h, 1,950K overall diffused light) with lights that consumed five times less power (metal lamps 70W/h, 3,000K cutoff light). The Seoul Metropolitan Government objected to our proposal, saying that the lower power consumption could be a safety issue at night. Light is not just about high power consumption. First, the light distribution characteristics of the luminaires and how they are presented must be examined. Currently, the biggest spoilers of Hangang River's night are the 400W/h high-powered, generalized diffuse lights installed in the park to ensure safety. It pours light upward toward the sky and downward into the river. This caused light to spill into the interior spaces of the surrounding residential apartments, causing light pollution that interfered with people's sleep.

We realized that using 400W/h of light in the park, which is equivalent to 1,000W/h of light at the Sangam-dong World Cup soccer stadium, was a problem in itself. We wanted to convince Seoul officials of the problem through real-world light testing. First, we demonstrated that light levels could be achieved with data through simulations of light calculation, and second, we verified our light design in a field experiment using a mobile silent generator on Seorae Island in the Banpo district of Hangang Park. The results of the field test confirmed that it is possible to secure sufficient road surface illumination, reduce the power required, and eliminate the glare and upward light caused by excessive light from the Hangang Park lights, which have been a problem in the past, to create a pleasant night view of the Hangang River.

노을의 품에 안긴 한강 | Hangang River embraced by sunset

우리의 빛 디자인 실험을 계기로 한강공원 조명의 소요 전력이 모두 낮아지게 됐다. 이런 획기적인 개선이 가능했던 이유는 책상에 앉아서 이론으로 세운 빛의 계획이 아니라 현장 중심의 빛 테스트를 통한 현장실험으로 문제를 해결하겠다는 의지 때문이었다. 우리의 이런 다각적인 빛의 접근법, 심야 시간까지 현장실험을 함께 수행한 서울시 관계자의 노고가 있었기에 가능했다.

지금도 빛에 대한 계획이 개별적으로 혹은 동시다발로 진행되고 있다. 그러나 야간경관처럼 주변 빛과의 상관관계에서 형성되는 밤의 빛 계획과 밤의 모습을 담는 빛 디자인에서는 현장에서의 빛 테스트 사례 공유가 중요하다. 실제 조도, 휘도 계산에 없었던 누수 되는 빛, 주변 건축물에 의해서 반사되는 빛 등을 실제로 보고 느껴보는 것이 중요하다. 이런 섬세한 부분을 종합적으로 이해할 때 빛의 콘셉트를 완성도 높게 만들어 낼 수 있다.

Our light design experiment resulted in a reduction in both the power requirements of the Hangang Park lights. This breakthrough was possible because of our willingness to solve problems through field experiments with field-oriented light testing, rather than theoretical light plans at a desk. Our multifaceted approach to light and the efforts of Seoul officials who worked late into the night to conduct field tests together made this possible.

Even now, light plans are being developed individually and in parallel. However, in the case of nighttime light planning and light design, which is shaped by the relationship with the surrounding light, such as a night view, it is important to share light test cases in the field. It is important to see and feel the actual light levels, the light leakage that wasn't accounted for in luminance calculations, and the light reflected by the surrounding architecture. A comprehensive understanding of these details will help you finalize your light concept.

서울시 여의도한강공원 빛 연출

한강르네상스 빛의 특화 디자인, 물빛이 흐르는 섬 여의도

여의도는 한강을 접한 섬으로, 국회와 언론사, 방송사, 금융회사 등이 위치하는 서울의 중심 업무공간으로, 고층 건축물이 밀집된 곳이다. 서울 한강르네상스 계획의 핵심 거점으로 새로운 변화를 알리는 상징적인 공간으로서 여의도한강공원이 새롭게 개발됐다. 우리는 사람과 자연이라는 관점에서 여의도한강공원 주변으로 흐르는 강물을 먼저 이해했고, 여의도한강공원에 물(Water)의 흐름을 빛으로 담고자 했다.

빛은 하늘에서 내려오고, 물은 대지를 따라 높은 곳에서 낮은 곳으로 흘러 한강이라는 이름으로 빛과 만난다. 빛은 어둠에 순응하여 일정 시간 머물다 새로운 자연 속으로 사라진다. 이런 자연의 흐름이 여의도한강공원 빛 연출의 중심 콘셉트이다. 여의도 고층빌딩에서 시작된 빛의 흐름을 여의도한강공원에 모아서 담는 것이다.

콘크리트 옹벽을 허물고 푸른 언덕 조성

여의도한강공원을 중심으로 새로운 공원을 조성하려는 대규모의 토목사업이 서울 곳곳에서 이루어졌다. 과거에 강물의 범람을 방지하기 위해서 설치해 놓은 거대한 콘크리트 옹벽을 철거하고 푸른 공원을 조성하는 토목사업이다. 한강에 접한 일부 지역은 땅을 파서 강의 흐름을 원활하게 조성하고, 강과 멀어진 여의도한강공원의 윗부분은 흙을 쌓아 자연스러운 언덕을 만들었다. 위에서 아래로 흐르는 경사면을 조성했다. 이유는 풍광에 대한 자연스러운 조망을 중요하게 여겼기 때문이다.

한강 주변 혹은 한강을 이용하는 유람선에서 한강을 바라볼 때 삭막한 회색빛 콘크리트 옹벽은 보이지 않고 넓게 펼쳐진 여의도한강공원의 푸른 잔디밭이 경사면으로 시원하게 조망되도록 조성한 것이다. 많은 시민이 여의도한강공원 잔디밭에서 다이내믹(Dynamic)한 서울의 생동감을 느낄 수 있는 녹색 공간이 조성된 것이다.

LIGHT PRODUCTION OF YEOUIDO HANGANG PARK IN SEOUL, KOREA

Yeouido, a Watery Island Specializing in Hangang River Renaissance Light Design

Yeouido is an island in the Hangang River, home to the National Assembly, media organizations, broadcasters, and financial firms, and is a central business district in Seoul with a high concentration of skyscrapers. As a key hub of Seoul's Hangang Renaissance Plan, Yeouido Hangang Park was newly developed as a symbolic space to signal the new transformation. We first understood the river flowing around Yeouido Hangang Park from the perspective of people and nature, and then tried to capture the flow of water in Yeouido Hangang Park with light.

Light descends from the sky, and water flows from high to low along the land, meeting with light in the name of the Hangang River. The light adapts to the darkness and stays for a period of time before disappearing into the new nature. This natural flow is the central concept behind the light production of Yeouido Hangang Park. The flow of light from the skyscrapers of Yeouido is collected and captured in Yeouido Hangang Park.

Tear Down Concrete Retaining Walls and Create Green Hills

A large-scale civil engineering project to create a new park centered on the Yeouido Hangang Park has been carried out all over Seoul. The massive concrete retaining walls that were built in the past to prevent the river from flooding were demolished and a green park was created. Some areas facing the Hangang River were dug up to create a smooth flow of the river, while the upper part of Yeouido Hangang Park, which is far from the river, was piled up to create a natural hill. We created a slope that flows from top to bottom. The reason for this is that the natural view of the landscape was important.

When looking at the Hangang River from the Hangang River area or from a cruise ship using the Hangang River, the dull gray concrete retaining wall is not visible, and the sprawling green lawn of Yeouido Hangang Park is visible from the slope. The green space was created so that many citizens can feel the vibrancy of dynamic Seoul from the lawn of Yeouido Hangang Park.

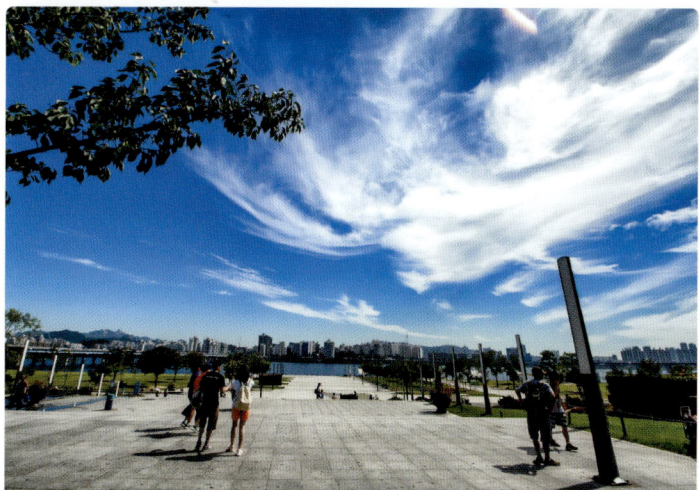

한강에서 가장 큰 섬, 밤섬과 여의도한강공원 | Largest Island, Bamseom and Yeoido Hangang Park, Hangang River

여의도는 작은 빛으로, 뺄셈으로 빛 디자인

우리는 한강공원에 빛을 디자인하기에 앞서, 어둠으로의 회복, 새로운 원칙과 절제를 내세웠다. 한강에서 가장 큰 섬 여의도의 가치를 절제된 작은 빛으로 표현하고 싶었고, 그 안에서 각별한 여의도 밤의 이야기를 들려주고 싶었다. 빛을 화려하게 덧입히기보다는 뺄셈을 통해 좀 더 아늑하게 여의도한강공원의 밤을 담아내겠다는 계획을 세웠다.

Yeouido Designs Light by Subtraction with Small Light

Before designing the light in Hangang Park, we proposed a return to darkness, a new principle and restraint. We wanted to express the value of Yeouido, the largest island in the Hangang River, with a small, understated light, and tell the story of a special Yeouido night. We planned to capture the night in Yeouido Hangang Park in a more cozy way through subtraction rather than adding more light.

여의도한강공원 빛의 디자인 | Light Design of Yeoido Hangang Park, 2008

건물에서 흘러내리는 빛줄기를 모아 '물의 풍광 (Waterscape)' 기획

여의도는 섬이다. 가장 높은 오피스 건축물에서 작은 빛의 물방울들이 반짝이고, 건축물을 따라 아래로 흘러내린다. 이러한 빛의 물방울이 대지의 물길을 따라 흘러서 한강의 큰 흐름에 합류한다. 이는 여의도 건물이 만들어 낸 빛의 흐름을 이미지로 그려낸 것이다.

Collecting Light Streams from Buildings to Create a 'Waterscape'

Yeouido, an island, showcases a stunning spectacle as tiny droplets of light shimmering from the tallest office buildings, cascading down the architecture. These droplets of light flow along the waterways of the land and join the larger flow of the Hangang River. This is an image of the flow of light created by the buildings in Yeouido.

여의도, 빛으로 수놓은 한강의 심장, 국제도시조명연맹(LUCI), CPL Award 2등 수상 | Lighting Urban Community International (LUCI), CPL "City.People.Light" Award, Second Prize, Heart of Hangang River embroidered with light, Yeouido Hangang Park, 2013

맑고 투명한 오피스 건축물의 빛이 청백색의 빛이라면 여의도한강공원을 적시는 빛은 따뜻하고 부드러운 빛이어야 했다. 새롭게 보여주기 위한 빛이 아니라 삶의 모습이 담긴 빛, 그리고 배려할 수 있는 빛, 이러한 빛의 연합으로 여의도한강공원의 밤을 빚어내는 물처럼 흐르는 빛 '물의 풍광(Waterscape)'이라는 빛의 콘셉트를 만들어 냈다.

If the light from the crystal-clear office buildings was blue and white, the light that bathed Yeouido Hangang Park had to be warm and soft. The combination of light, which is not a light to show something new, but a light that contains life and a light that can be considered, created the light concept of 'Waterscape,' a light that flows like water to form the night of Yeouido Hangang Park.

물방울의 빛 | The light of a water droplet

밝고 따뜻한 빛이 흐르는 여의도한강공원의 빛

그 흐름의 첫째는 공원을 진입할 때 밝게 빛나는 계단 옆으로 나열되어 출구와 입구를 알려주는 수직으로 늘어뜨린 빛의 열주(列柱)다. 빛의 열주는 고층 건물에서 흘러내린 빛방울을 형상화한 것이다. 이를 고스란히 공원 입구로 옮겨서 빛의 수직 구조로 디자인했다.

입구를 지나면 빛의 긴 선형을 만난다. 여의도한강공원에서 가장 밝게 빛나는 빛, 자전거도로의 빛이다. 빠른 속도로 질주하는 자전거와 인라인 타는 사람들의 안전을 확보해 주기 위한 빛으로, 자전거도로 노면에 균일한 밝기를 확보함으로써 시원하고 경쾌한 빛을 먼저 만난다.

빛이 아늑하게 머문다

여의도한강공원의 산책로를 따라 한강으로 가까이 다가가면 다가갈수록 조금은 어둡고 산뜻한 빛을 만나게 된다. 이는 강변으로 다가가면서 점차 조도를 낮추어서 편안함을 느낄 수 있도록 빛을 배려했기 때문이다. 이러한 방법을 사용하기 위하여 자연녹지가 있는 공간의 빛 밝기를 낮추고 강물로 가는 빛을 차단하기 위해 빛가림막을 적용했다. 또한 여의도한강공원 선착장과 사람들이 많이 모이는 공간에는 조도를 높여서 밝고 활기가 넘치도록 빛을 디자인했다.

공원의 빛은 모두 같은 것이 아니다. 한강 수변공간의 특성과 이곳을 찾는 사람들의 행동 패턴을 고려하여 빛에 리듬감을 부여하여 섬세하게 디자인한 것이다. 빛 연출은 빛의 색감을 고려하여 공간별로 조금씩 다르게 디자인했다. 진입부와 주차장은 맑고 산뜻한 백색(4,000K)으로, 공원의 산책로는 부드럽고 아늑한 백색(3,000K)을 사용해서 서울의 중심 공간으로서의 독특한 공원 모습으로 빚어냈다. 그리고 그 속에 아주 작고 맑고 투명한 물빛을 드러내고자 푸른빛을 여의도한강공원 캐스케이드(Cascade) 수변 공간 바닥에 광섬유(Fiber Optic)를 사용하여 물빛을 강조하도록 디자인했다.

여의도한강공원의 빛을 디자인하면서 다양한 빛의 가치를 새롭게 느낄 수 있었다. 공원에서의 빛은 일률적으로 밝고 선명하게 밝히는 것만 중요한 게 아니라, 장소의 특성과 이용자의 특성에 따라 빛을 다르게 설정해야 한다는 점이다. 특히 공간을 이용하는 사람들에 대한 배려가 중요했다. 이렇게, 좋은 빛을 디자인하려면 먼저 공간과 장소를 이해하고, 그 공간을 이용하는 사람들의 마음(Soul)을 담아내는 기획이 필요하다.

The Bright, Warm Glow of Yeouido Hangang Park

The first of these flows are the vertical colonnades of light that flank the illuminated staircase as you enter the park, marking exits and entrances. The colonnades are meant to resemble drops of light falling from skyscrapers. This was carried over to the park's entrance, which was designed as a vertical structure of light.

After passing through the entrance, visitors encounter a long linear line of light. The brightest light in Yeouido Hangang Park is the bicycle path. This light is meant to ensure the safety of bicyclists and inline riders who are traveling at high speeds, and the uniform brightness of the bicycle road surface creates a cool and cheerful light.

Light Stays Cozy

As you approach the Hangang River along the promenade of Yeouido Hangang Park, you will notice that the light is slightly dimmer and more refreshing as you approach. This is because the light level is gradually lowered as you approach the riverfront to make you feel more comfortable. To do this, the brightness of the natural green spaces was reduced and light shields were applied to block the light from the river, while the light levels were increased in the marina and other areas where people gathered to create a bright and lively atmosphere.

Not all parks are created equal. It was delicately designed to give a sense of rhythm to the light by considering the characteristics of the Hangang waterfront space and the behavior patterns of people visiting the park. The light production was designed slightly differently for each space, taking into account the color of the light. The entrance and parking lot are crisp white (4,000K), while the park's walkways are soft and cozy white (3,000K), giving the park a unique look as a central space in Seoul. To reveal the tiny, crystal-clear water, the blue light was designed to highlight the water using fiber optics at the bottom of the Cascade waterfront space in Yeouido Hangang Park.

While designing the light for Yeouido Hangang Park, I was able to realize the value of various types of light. It is not only important that the light in the park be uniformly bright and clear, but that it should be set differently according to the characteristics of the place and the characteristics of the users. In particular, it was important to consider the people who use the space. In this way, to design good light, it is necessary to first understand the space and the place, and to plan to capture the soul of the people who use the space.

빛의 언덕, 여의도한강공원 캐스케이드와 프롬나드 | Hill of Light, Cascade and Promenade in Yeouido Hangang Park

빛과 자연의 조화, 여의도한강공원 프롬나드 | Harmony of light and nature, Yeouido Hangang Park Promenade

서울시 난지한강공원 빛 연출

LIGHT PRODUCTION OF NANJI HANGANG PARK IN SEOUL, KOREA

노을빛, 별빛, 달빛이 아름다운 난지한강공원

난지한강공원은 한강 서쪽에 있는 자연생태 보전지역이다. 주변에는 노을공원과 하늘공원, 평화공원인 생태환경공원이 인접해 있다. 자연생태공원이자 수변 공간으로서의 장소적 특성을 고려하여 서울에서 가장 밤다운, 밤의 멋과 밤의 낭만을 담아내는 빛을 디자인했다. 수변 산책로는 자연스러운 리듬감이 있는 빛으로 연출하고, 난지한강공원 일부 공간에서 낭만적인 밤을 느낄 수 있는 빛을 창출했다.

노을은 그리움을 빚어낸다

노을공원과 하늘공원을 따라 흐르는 빛은 난지한강공원 산책로에서 어둠 속으로 일부가 사라지고, 한강과 가까운 산책로에서는 수면 위에 잠시 머물다 한강의 밤 속으로 사라진다. 수면 위에 빛나는 별빛이 쏟아지는 모습으로 공원의 빛을 디자인했다.

강변 고속도로와 접한 부분에서는 빛이 도로로 유입되는 것을 차단했고, 성산대교 등 한강교량에 인접한 산책로에서는 주변 공간이 밝게 빛나도록 디자인했다. 생태습지원에서는 어둠 속에서 최소한의 빛으로 이곳을 이용하는 사람들의 안전을 확보할 수 있도록 전체적으로 2,800K~3,000K의 따뜻한 빛으로 연출하고, 공원 보안등의 높이도 휴먼스케일을 고려하여 보행자에게 심리적 안정감이 느껴지도록 빛을 새롭게 만들어 냈다.

Nanji Hangang Park with Beautiful Sunset, Starlight and Moonlight

Nanji Hangang Park is a natural ecological conservation area on the west side of the Hangang River. The park is surrounded by Noeul Park, Sky Park, and Peace Park. Considering the characteristics of the site as a natural ecological park and a waterfront space, we designed the lighting to capture the nighttime charm and romance of Seoul's nightlife. The waterfront promenade is illuminated with a natural rhythmic light, and in some areas of Nanji Hangang Park, we created a romantic nighttime atmosphere.

Sunset Creates Nostalgia

The light along Noeul Park and Sky Park partially disappears into the darkness on the Nanji Hangang Park promenade, and on the promenade close to the Hangang River, it lingers briefly on the surface of the water before disappearing into the night of the Hangang River. The park's light is designed to resemble starlight shining on the surface of the water.

In areas bordering the riverside highway, we blocked light from entering the roadway, and on trails adjacent to Hangang bridges such as Seongsan Bridge, we designed the surrounding space to shine brightly. In the ecological wetland area, the overall warm light of 2,800K to 3,000K is used to ensure the safety of people using the area with minimal light in the dark, and the height of the park security lights is also human-scale to create a new light that gives pedestrians a sense of psychological security.

난지한강공원 주간 조감도 | Daytime aerial view of Nanji Hangang Park

난지한강공원 배치도 | Site plan of Nanji Hangang Park

난지한강공원은 서울의 대표적인 생태공간

난지한강공원에는 사람도 있지만, 사람만큼이나 소중한 자연적인 생태공간이 널려있다. 여의도한강공원은 사람들을 위한 빛의 계획이었다면 이곳은 하늘의 별빛과 달빛을 위해 조성된 자연 그대로의 모습을 바라볼 수 있는 빛의 공간이다.

  우리는 공간을 디자인하면서 '다르다'라는 기본적인 생각을 가끔은 놓치기도 한다. 모두가 유사하다고 생각하는 것이다. 그래서 나는 자신을 향해 자주 질문한다. 이 공간에 꼭 이 빛이 적합한 것인지, 만약에 내 생각이 잘못되었다면 과감하게 현장을 다시 찾아가서 질문에 대한 답을 찾는다. 그리고 그 안에 담을 독특한 빛을 다시 생각한다. 우리는 그곳에 적합한 빛을 꿈꾸는 것이지 조명시설물을 설치하려는 목적이 아니니까. 빛이 만들어 주는 진정한 가치를 탐색하려는 노력이 필요하다. 주변 환경이 어둡다면 작은 빛으로도 충분히 그 공간의 가치를 드러낼 수 있다는 믿음으로 난지한강공원만의 독특한 밤을 창출해 냈다. 난지한강공원은 아이러니하게도 아무것도 하지 않는 빛 디자인, 밤의 본래 모습으로 돌려놓는 작업 중심으로 진행했다.

Nanji Hangang Park is Seoul's representative Ecological Space

Nanji Hangang Park has people, but it is also filled with natural ecological spaces that are just as important as people. While Yeouido Hangang Park was a plan of light for people, this is a space of light where you can gaze at the pristine nature created for starlight and moonlight in the sky.

  When we design spaces, we sometimes lose sight of the basic idea 'different'. We think everyone is similar. So I often ask myself the question. If I'm wrong, I boldly go back to the site to find the answer to the question and rethink the unique light that I want to put in there. We're dreaming of the right light for the space, not trying to install a fixture. It is about exploring the true value of light. We believe that if the surrounding environment is dark, even a small light can reveal the value of the space, so we created a unique night in Nanji Hangang Park. The design of Nanji Hangang Park centered on designing a light that ironically does nothing, returning the night to its original state.

난지한강공원 야간 조감도 | Nighttime aerial view of Nanji Hangang Park, 2009

난지한강공원 자전거도로와 강변수영장 | Nanji Hangang River Park bicycle path and riverside swimming pool.

자연을 담다:
호수와 대나무 숲 그리고 라운지 공간

EMBRACE NATURE:
LAKE, BAMBOO FOREST AND LOUNGE AREA

경기도 광교신도시 호수공원 빛 연출

LIGHT PRODUCTION OF GWANGGYO NEW CITY LAKE PARK, GYEONGGI-DO

전라남도 담양 죽녹원 빛 연출

LIGHT PRODUCTION OF JUKNOKWON GARDEN IN DAMYANG, JEOLLANAM-DO

인천국제공항 아시아나항공 라운지 빛 연출

LIGHT PRODUCTION OF ASIANA AIRLINES LOUNGE AT INCHEON INTERNATIONAL AIRPORT

나는 꿈을 꾼다. 이곳에서 스며드는 감성을
그리고 깊은 감사함을 느낀다.
나의 감성, 나의 빛이, 이곳에 고스란히 담겨 있음을.

I dream. The emotions that seep in from this place
and I feel a deep sense of gratitude.
My emotions, my light, are fully captured here.

경기도 광교신도시 호수공원 빛 연출

LIGHT PRODUCTION OF GWANGYO NEW CITY LAKE PARK, GYEONGGI-DO

풍광명미(風光明媚), 맑은 물가를 거닐며 밤이 아름다운 호수를 만나다.

경기도 수원시와 용인시 사이에 경기도청이 이전된 광교신도시가 있다. 광교신도시 한가운데 맑은 물을 담고 있는 원천호수와 신대호수 두 호수가 있다. 이곳을 광교호수공원이라고 한다. 광교호수공원은 2010년 6월부터 2013년 봄까지 호수 환경 개선 사업을 시작하여 3년 동안 호수공원 조성 사업이 진행되었다.

두 호수는 단순히 물을 담아놓은 저수지 개념에서 벗어나 신도시 중심 수변공간, 휴식 공간 조성이라는 관점에서 국제공모전으로 조경디자인 설계가 진행되었다. 새롭게 선정된 광교호수공원 조경디자인 작품의 콘셉트는 '어반 소프트 파워(Urban Soft Power)'로서 광교산과 광교호수 본래 모습을 최대한 보존하고, 이곳에 새로운 도시 문화를 담아낸 수변공원을 조성한다는 디자인 방향을 잡았다.

광교호수공원 빛 디자인 계획은 주변 광교산과 조화를 이룰 수 있는 밤의 풍광 창출을 목표로 빛 연출을 디자인했다. 빛의 콘셉트는 풍광명미(風光明媚)*다. '맑은 물가를 거닐며 밤이 아름다운 호수를 만나다.'라는 호수의 원형 그대로의 모습을 지켜 내고자 했던 콘셉트다. 밝음을 즐기는 곳과 밤의 어둠을 담는 낭만적인 공간으로 빛을 기획했다.

*풍광명미(風光明媚): 산수의 경치가 맑고 아름다움

Stroll Along the Crystal-Clear Waters and Discover the Lake at Night

Gwanggyo New City, where the Gyeonggi-do provincial government was relocated, is located between Suwon and Yongin. In the center of Gwanggyo New City, there are two lakes, Wonchon Lake and Shindae Lake, which contain clear water. This place is called Gwanggyo Lake Park, which started a project to improve the lake environment from June 2010 to spring 2013, and the project to create a lake park lasted for three years.

The landscape design of the two lakes was conducted through an international competition from the perspective of creating a waterfront space and resting area at the center of the new city, rather than just a reservoir for water. The concept of the newly selected Gwanggyo Lake Park landscape design work is 'Urban Soft Power', and the design direction is to preserve the original appearance of Gwanggyo Mountain and Gwanggyo Lake as much as possible and create a waterfront park that reflects a new urban culture.

The light design plan for Gwanggyo Lake Park aims to create a nighttime landscape that harmonizes with the surrounding Gwanggyo Mountain. The concept of the light is called "Punggwang Myeongmi"*. The concept was to preserve the original form of the lake, which is 'walking along the clear water and encountering the beautiful lake at night'. The light was planned as a place to enjoy the brightness and a romantic space to capture the darkness of the night.

*Punggwang Myeongmi: The clarity and beauty of the scenery of mountain and water

광교호수공원 빛의 마스터플랜
Light Master Plan at Gwanggyo Lake Park, 2013

광교호수공원 배치도 | Gwanggyo Lake Park Layout Plan

낮과 밤이 다른 빛의 기획

원천호수와 신대호수 공원의 토속적인 자연경관을 유지하기 위해 조명시설물 설치를 최소화하고, 일부 조명시설은 조경 시설물과 일체형으로 빛을 만들어서 공간을 연출했다. 수변 산책로는 난간 내부에 빛을 설치해서 낮에는 조명시설이 보이지 않도록 하고, 일몰 후 시간의 변화에 따라 파스텔 색감으로 변화되어 시간 변화를 시각적으로 느낄 수 있도록 차별화된 빛을 디자인했다.

Day and Night Light Planning

In order to maintain the indigenous natural landscape of Woncheon Lake and Shindae Lake Park, the installation of lighting facilities was minimized, and some lighting facilities were integrated with landscape facilities to create a space. For the waterfront promenade, lights were installed inside the railings to prevent the lighting facilities from being visible during the day, and the lighting was designed to be differentiated by changing to pastel colors after sunset to visually feel the change of time.

*컷오프 조명(Cutoff lighting): 빛 공해를 예방하는 방법으로, 조명과 지면이 동일하게 수평이 되도록 설치하여 빛이 하늘로 올라가지 않게 만드는 조명 방법이다.

*Cutoff lighting: A method of lighting that prevents light pollution by installing lights so that they are level with the ground so that light does not rise into the sky.

보행자의 안전을 확보하기 위한 보안등은 3가지 디자인 콘셉트로 계획하였다. 첫째, 경기도청과 호텔, 컨벤션센터 등 도시 시설과 인접한 원천호수 주변 지역의 보안등은 선형의 빛이라는 개념으로 보안등 형태를 비스듬한 곡선으로 디자인해서 자연스럽게 사람들의 시선이 호수로 향하도록 만들었다. 둘째, '재미있는 밭'이 있는 숲속 산책로는 따스한 빛의 콘셉트로 산책로 양방향 확산형 조명 방법을 적용하여 시민들에게 '무서운 숲'이 아니라 '아늑한 숲'이 되도록 조성했다. 셋째, 자연의 풍미를 존중하는 신대호수는 자연의 색감을 나타낸다는 방향으로 습지와 버드나무 숲이 어우러진 자연형 수변공간이 드러나도록 설계했다. 이곳은 수변에 접한 보행 공간이어서 목재 형태의 질감을 사용하여 폴을 만들고, 일부 빛이 하늘로 상향되지 않도록 컷오프 조명(Cutoff lighting)*으로 빛을 디자인했다.

The security lights to secure pedestrian safety were planned in three design concepts. First, the security lights in the area around Woncheon Lake, which is adjacent to city facilities such as the Gyeonggi-do Provincial Government, hotels, and convention centers, were designed with the concept of linear light, and the shape of the security lights was designed as an oblique curve to naturally draw people's attention to the lake. Second, the forest walkway with the 'fun field' was designed with the concept of warm light, applying a diffuse lighting method in both directions of the walkway to create a 'cozy forest' rather than a 'scary forest' for citizens. Third, Shindae Lake, which respects the flavor of nature, was designed to reveal a natural waterfront space with wetlands and willow forests in the direction of representing the colors of nature. As this is a walking space facing the water, we used wood-like textures to create poles, and designed the lighting with cutoff lighting* to prevent some light from going up to the sky

광교호수공원 원천호수의 선형을 만들어 내는 빛, 어반레비 | Light that creates the line of the Woncheon Lake in Gwanggyo Lake Park, Urban Levy

시간마다 다르게 움직이는 빛, 신비로운 빛 선사

시민들의 문화공간인 원천호수 수변 산책로를 "시간에 따라 달라지는 빛으로 색다른 분위기를 만든다."라는 새로운 발상으로 빛을 계획했다. 일몰 이후 시간의 흐름에 따라 다양한 색깔과 함께 독특한 분위기로 바뀌는 역동적인 빛을 원천호수에 담아 보았다. 오후 7시 정각에는 수변 데크 산책로의 빛이 푸르시안 블루(Prussian Blue Color), 8시 정각에는 레드 퍼플(Red Purple Color), 9시 정각에는 버밀리언 컬러(Vermilion Color), 10시 정각에는 브라운 컬러(Brown Color)로 변화하도록 빛을 디자인했다. 이러한 빛의 색감은 시간에 따라 변화하여 밤에 호수 주변을 산책하는 시민들에게 원천호수만의 물이 주는 빛의 감성을 느끼는 시간을 만들어 주었다.

Light Moves at Different Times of Day, Creating a Mysterious Glow

The promenade along the waterfront of Woncheon Lake, a cultural space for citizens, was planned with a new idea of "creating a different atmosphere with light that changes with time". After sunset, Woncheon Lake is illuminated with a dynamic light that changes into a unique atmosphere with various colors as time passes. At 7 p.m., the light on the waterfront deck walkway changes to Prussian blue, at 8 p.m. to red purple, at 9 p.m. to Vermilion, and at 10 p.m. to Brown. These colors change over time, giving people walking around the lake at night the opportunity to feel the light emotions of the water of Woncheon Lake.

19:00 Prussian Blue Color

20:00 Red Purple Color

21:00 Vermilion Color

빛의 프롬나드, 광교호수공원 | Light Promenade, Gwanggyo Lake Park

빛이 흐르는 광교호수공원 어반레비 | Urban Levy, where light flows, Gwanggyo Lake Park

신대호수와 원천호수의 대비되는 빛 연출

광교 호수 중 비교적 자연경관을 고스란히 담고 있는 곳이 신대호수이다. 바로 옆에 위치한 원천호수와 반대되는 빛으로 연출했다. 나무가 무성한 숲속 산책로 공간에 나무 그늘로 어두운 곳은 일부 수목을 업라이트(Up Light)로 비추어서 수목에 반사된 빛이 주변을 밝혀 안전한 산책 공간으로 조성했다. 보행로 스텝 조명은 아래에서 옆으로 향하게 연출된 따뜻한 빛을 만들어 보행자에게 밤의 정감을 느낄 수 있게 했다. '문라이트(Moon light)'로 빛을 디자인하여, 달을 닮은 조명을 신대호수 수면에 띄웠다. 이로써 낮과 다른 달밤 호수의 새로운 공간이 창출되었다.

소리도 디자인하고 싶은 빛 디자이너

나는 빛의 디자이너이자 빛 연출가로 단순히 빛을 디자인하는 것이 아니라 공간에 맞는 빛의 이미지를 기획하고, 그 안에 살아있는 빛은 물론이고, 빛과 어울리는 소리를 디자인하여 밤의 공간 분위기를 낮과 다른 이미지로 연출하기 위해 노력했다. 광교호수공원을 방문하는 많은 사람이 수변공간만이 가지고 있는 물이라는 이미지와 편안함을 느낄 수 있는 공간으로 만들어 주고 싶었다. 그래서 최대한 빛이 인위적이지 않도록 자연과 호흡하는 공간이 되어주기를 소망하면서 공간별로 빛을 다르게 배치하고, 일부 물너미 공간에서는 소리도 함께 디자인하여 빛 연출 작품을 만들어 냈다.

광교호수공원 야간경관 디자인은 국토교통부가 주최하는 「2014 대한민국 경관대상」에서 야간경관으로 대상(국토교통부 장관상)을 받았다.

Creating contrasting light between the Shindae Lake and Woncheon Lake

Among the lakes in Gwanggyo, Shindae Lake is the one that has a relatively natural landscape. The light is opposite to that of the neighboring Woncheon Lake. In the pathway through the shadowy woods, some trees were lit from below, and the light reflected from them brightened the surroundings, creating a safe walking environment. The walkway step lights created a warm glow directed downward and sideways, giving pedestrians a sense of nighttime. We designed the light as a 'moon light', floating a light that resembles the moon on the surface of Shindae Lake. This created a new space in the lake at night that was different from the daytime.

Light Designers Who also Want to Design Sound

As a light designer and light director, I do not simply design light, but I plan the image of light that fits the space, design the light that lives in it, and design the sound that goes with the light, so that the atmosphere of the space at night is different from that of the day. I wanted to create a space where many people who visit Gwanggyo Lake Park can feel comfortable with the image of water that only a waterfront space can have. Therefore, I wanted to make it a space that breathes with nature so that the light is not artificial as much as possible, so I arranged the light differently for each space and designed sounds for some of the waterfront spaces to create light production works.

The night view design of Gwanggyo Lake Park won the Grand Prize (Minister of Land, Infrastructure, and Transport Award) for night view at the 2014 Korea Landscape Awards organized by the Ministry of Land, Infrastructure, and Transport.

전라남도 담양 죽녹원 빛 연출

LIGHT PRODUCTION OF JUKNOKWON GARDEN IN DAMYANG, JEOLLANAM-DO

대나무 숲 죽녹원에 달이 뜨면 한 폭의 수묵화가 펼쳐진다

담양의 죽녹원(竹綠苑)은 성안산 일대에 약 16만㎡의 울창한 대숲으로, 2003년 5월에 개원했다. 울창한 대나무 숲과 2.2Km의 대나무 숲길을 직접 체험할 수 있는 산책로로 구성되어 있다. 운수대통길, 죽마고우길, 철학자의 길 등 8가지 주제의 분위기가 다른 산책로가 조성되었는데, 죽녹원 전망대로부터 산책로가 시작된다.

생태전시관, 인공폭포, 생태연못, 야외공연장이 있으며, 밤에도 산책할 수 있도록 대숲에 조명시설이 되어 있다. 2017년 기준으로 내외국인 관광객 연 140만 명이 찾는 인기 높은 테마공원으로, 같은 해 서울 창덕궁 입장객이 180만 명이니 입장객이 얼마나 많은지 가늠할 수 있다.

죽녹원을 거닐다 보면 사람을 차분하게 만드는 매력적인 분위기에 빠져들고, 대나무와 댓잎이 자아내는 청량한 정취도 함께 느낄 수 있다. 그러나 죽녹원은 밤이 되면 대나무 숲 그늘로 어둠이 짙어지며, 댓잎 부딪히는 소리까지 더해져 산책로가 다소 무서운 공간이 되기도 한다. 우리는 죽녹원 자연경관을 지켜내면서 대숲의 운치가 있는 빛 환경 조성을 위한 빛 디자인 작업을 진행했다.

When the moon rises in the bamboo forest, Juknokwon, a black and white painting unfolds

Damyang's Juknokwon Garden is a 160,000-square-meter bamboo forest located in Seongansan mountain, and opened in May 2003. It consists of a dense bamboo forest and a 2.2-kilometer bamboo forest trail where visitors can experience the bamboo forest firsthand. There are eight trails with different themes, including the Unsu-daetonggil Trail, the Jukmagougil Trail, and the Cheolhagjauigil Trail, and the trail starts from the Juknokwon Garden Observatory.

There is an ecological exhibition center, an artificial waterfall, an ecological pond, and an outdoor performance center. The bamboo forest is also illuminated for nighttime strolls. As of 2017, this popular theme park attracts 1.4 million domestic and foreign tourists annually, which is significantly higher compared to the 1.8 million visitors to Changdeokgung Palace in Seoul in the same year.

As you stroll through Juknokwon Garden, you'll find yourself immersed in a charming atmosphere that makes you feel calm, and you can also feel the refreshing atmosphere of bamboo and bamboo leaves. However, at night, Juknokwon Garden is darkened by the shade of the bamboo forest, and the sound of the leaves crashing against each other can make the walkway a somewhat frightening place. We worked on a light design to create a light environment with the atmosphere of a bamboo forest while preserving the natural landscape of Juknokwon Garden.

담양군 죽녹원의 대나무 | Bamboo in Juknokwon, Damyang-gun

자연 미학을 담는 빛 연출

죽녹원의 빛 디자인 조성 시에 가장 먼저 염두에 둔 것은 자연 그대로의 미학을 담기 위한 빛 연출에 대한 고민이었다. 나는 죽녹원 대나무 숲 자체가 하나의 빛이 되어야 한다고 생각했고 빛의 콘셉트는 먹의 농담(濃淡)으로 빚어지는 수묵화(水墨畵)였다. 대나무와 대나무 사이로 스며 나오는 햇살이 주는 따스함을 죽림에 투영하여 편안한 수묵화의 산책로를 빚어낸 것이다.

    빛은 대나무의 직선과 부드러운 곡선이 빚어내는 형상을 산책로에 담아서 죽녹원에 한 폭의 수묵화가 그려지기를 바라는 마음으로 빛을 디자인했다. 마치 인상파의 화가들이 빛으로 표정을 묘사하고, 그림자의 인상을 캔버스에 채색한 것처럼 빛과 그림자가 빚어낸 대숲의 풍광을 연출하고자 했다.

    나는 빛의 디자인과 빛 연출 현장 감리인 총감독 컨설팅을 통해서 이곳에다 한 폭의 수묵화를 빛으로 그려냈다. 그림을 그리기 위한 한지(韓紙)가 죽녹원의 산책로가 되고, 대나무 숲 사이로 새어 나오는 대나무의 그림자가 수묵화의 대나무 선형이 된다. 빛의 붓으로 잎을 그려내니 산뜻한 수묵화가 죽녹원 산책로에 펼쳐졌다.

Create Light that Captures Natural Aesthetics

The first thing I kept in mind when creating the light design for Juknokwon was how to create light to capture the natural aesthetic. I believe that the bamboo forest itself should harmonize into one light, and the concept of the light was an ink-and-wash painting. The warmth of the bamboo and the sunlight seeping through the bamboo was projected onto the bamboo forest to create a relaxing ink painting walkway.

    The light was designed with the hope that the straight lines and gentle curves of the bamboo would reflect in the walkway, and that the bamboo gardens would become the center of an ink-and-wash painting. I wanted to create a bamboo forest landscape of light and shadow, just as impressionist painters used light to depict facial expressions and painted the impressions of shadows on canvas.

    Through consulting with the light design and the general director who supervised the light production site, I wanted to create a series of ink-and-wash paintings with light. The Hanji (Korean paper) for drawing becomes the walking path of the bamboo grove, and the shadows of the bamboo leaking through the bamboo grove become the bamboo linear lines of the ink-and-wash painting. As I painted the leaves with a light brush, a refreshing ink-and-wash painting spread out on the bamboo walkway.

빛의 디자인, 죽녹원 | Light Design, Juknokwon, 2013

죽녹원의 빛은 여백의 무색이다

죽녹원 대나무 숲에 수묵화를 펼쳐놓기 위해서 빛의 디자인 원칙 세 가지를 설정했다.

첫째, 빛은 있지만 빛이 없는 무광(無光)의 여백(餘白)을 만든다. 죽녹원 대나무 숲 경관을 훼손하지 않는다는 첫 약속이 무광이었다. 무광이란 '빛은 있되 빛이 없다.'라는 뜻이다. 이는 죽녹원의 밤 자체를 지키고 싶다는 마음으로 밤을 밝히기 위해서 설치되는 인공적인 조명시설물에 대해 낮과 밤 모든 탐방객의 시각에 인지되지 않는 빛을 만들어 나가는 것이다. 빛은 있는데 빛이 시각에 들어오지 않으니 죽녹원의 산책로에서 만나는 빛은 자연스럽게 대나무의 선형을 나타내고, 대나무숲은 평온한 공간이 된다.

둘째, 평온함을 담는 무심(無心)의 공간을 만든다. 죽녹원을 찾아오는 방문객을 위해 마음의 안정을 지켜주려는 약속이 무심이다. 죽녹원 대나무 숲의 산책로는 날씨가 흐린 날은 낮에도 어둡다. 이런 날 대나무 숲은 고요함을 넘어 음산한 느낌마저 든다. 당연히 밤에는 더 무섭다. 그래서 죽녹원을 찾는 누구도 불안하지 않도록 먼저 따뜻한 빛 속에서 산책할 수 있는 공간을 만들고자 했다. 산책로 대숲 좁은 공간에 새 지저귀는 소리를 배경음악으로 대나무 숲에서 행복감을 느끼게 하고, 수직으로 내려오는 대나무 빛이 더해져 죽녹원 고유의 평온한 공간이 조성되게 했다.

셋째, 빛으로 한 폭의 수묵화처럼 한지(韓紙)와 같은 무색의 공간을 펼쳐놓았다. 죽녹원은 낮에 아무것도 없는 대나무 숲으로만 존재한다. 우리는 이곳에 다양한 색감의 빛을 사용하지 않았다. 그리고 죽녹원만의 공간 미학을 담아내고자 수묵화 먹의 농담과 닮은 빛으로 죽녹원 대나무 숲을 연출해 냈다.

죽녹원에는 LED 조명의 현란한 컬러가 단 한 자락도 없다. 대나무 그대로의 빛과 그림자, 여백에 먹빛만 존재한다.

The Light of Juknokwon Garden is Colorless in the Blank Space.

I set three light design principles for the ink-and-wash painting in the bamboo forest of Juknokwon Garden.

First, create a matte white space where there is light but no light. Matte was the first promise we made to the bamboo forest landscape of Juknokwon Garden. It means 'there is light, but there is no light.' This means that the artificial light fixtures that are installed to illuminate the night in the hope of preserving the night itself will be imperceptible to all visitors, day and night. Since there is light, but the light is not visible, the light encountered on the trails of Juknokwon Garden naturally represents the linearity of bamboo, and the bamboo forest becomes a calm space.

Second, it creates a space of mindlessness that reflects tranquility. This is the promise of peace of mind for visitors to Juknokwon Garden. The bamboo forest walkway in Juknokwon Garden is dark even during the day when the weather is cloudy. On such days, the bamboo forest is not only silent but also eerie. Naturally, it is even scarier at night. So we wanted to create a space where visitors could walk in the warm light, therefore no one would feel uneasy. The chirping of birds in the narrow space of the bamboo forest on the promenade creates a sense of happiness in the bamboo forest, and the vertical bamboo light creates a tranquil space unique to Juknokwon Garden.

Third, the light spreads out a colorless space like Hanji, like an ink-and-wash painting. Juknokwon Garden exists only as a bamboo forest with nothing in the daytime. We did not use different colors of light here, and to capture the unique spatial aesthetic of Juknokwon Garden, we created the bamboo forest with light that resembles the shades of gray in an ink-and-wash painting.

There is not a single flashy color of LED lights in Juknokwon Garden. There is only the light and shadow of the bamboo and the ink color in the margins.

대나무의 그림자가 빚어내는 아늑한 산책로

죽녹원 비탈진 산책로는 자연스러운 빛 연출을 위해서 일률적인 업라이트(Up Light)에 의한 조명을 지양하고 대나무가 서 있는 경사도와 대나무의 높이를 고려하여 어떤 것은 대나무 줄기를 빛으로 연출하고 또 다른 것은 대나무 잎을 빛으로 연출해서 대나무의 수직적인 선형의 아름다움이 리듬감 있게 나타날 수 있도록 빛을 디자인했다. 또한, 숨어 있는 빛은 각각의 현장 테스트를 통해 설치할 위치를 설정하고 모든 빛이 완성되었을 때 또다시 빛의 각도를 하나하나 조정해서 대나무의 선형이 주변 공간 대나무의 울창한 정도에 따라 각기 다른 빛과 그림자를 빚어내도록 다양한 빛을 창출했다.

밤에 펼쳐지는 수묵화 속을 거닐다

해가 지고 어둠 속에서 빛이 서서히 살아날 때, 빛의 붓으로 그려지는 또 하나의 신비한 자연을 만나게 된다. 바로 숨어 있는 빛이 만들어 내는 부드러운 대나무 선형의 그림자. 죽녹원 산책로에 밤의 어둠에서 수묵화 같은 자연 선형인 대숲 먹빛 그림자의 농담(濃淡)이 수묵화처럼 번져간다.

A Cozy Walk in the Shadow of the Bamboo Trees

For the sloping walkway of Juknokwon Garden, we avoided lighting with uniform uplight to create a natural light, and designed the light in such a way that the vertical linear beauty of the bamboo can appear rhythmically by considering the slope and height of the bamboo, with some lights illuminating the bamboo stems and others illuminating the bamboo leaves. The concealed lights were also field-tested for each installation, and when all the lights were completed, the angles of the lights were adjusted again, one by one, to create a variety of light so that the linearity of the bamboo would create different light and shadow depending on how dense the bamboo was in the surrounding space.

Stroll through an Ink-and-Wash Painting Unfolding at Night

As the sun sets and the light slowly emerges from the darkness, we encounter another mystery of nature painted with the brush of light. It is the shadows of soft bamboo linear shapes created by the hidden light. In the stillness of the night, the bamboo forest casts graceful, gray shadows, evoking the elegance of ink-and-wash paintings adorning the Juknokwon Forest Trail.

산책로의 빛.
빛이 중첩되면서 만들어지는 대나무 그림자는 죽녹원을 더욱 아늑하게 만들어 낸다.

The light of the walking path.
As the light overlaps, the bamboo shadows created make Juknokwon Garden even cozier.

인천국제공항 아시아나항공 라운지 빛 연출

한국적인 전통미로 품격있는 공간 창출
인천국제공항 제1청사에 있던 대한항공의 라운지가 제2청사로 이전함에 따라, 이곳에 아시아나항공의 퍼스트, 비즈니스, 스타 라운지 공간이 새롭게 조성되었다. 아시아나항공의 기대는 한국적인 전통미로 아시아다운 품격을 갖춘 공간 창출이었다. 우리는 아시아나항공의 빛 연출 디자인에 앞서, 기존에 항공사 라운지에 조성해 놓은 실내조명과 인천국제공항공사 실내조명들을 두루 분석했다. 기존에 설계된 내용은 전체적으로 획일화되어 창백한 느낌의 조명으로 설계가 되어 있었다. 인테리어 마감 재료의 특성을 나타내지 못하고 단순히 조명 기능으로만 설치한 사실을 알았다.

정성스럽게 고객을 맞이하려는 따뜻한 분위기 창출
우리는 고객을 '마음을 가다듬어 정성스럽게 맞이하려는 정심성의(正心誠意)' 아시아나항공의 마음을 빛으로 담아내고 싶었다. 오랜 시간의 비행을 기다리는 초조한 사람들에게 정감을 느끼도록 하겠다는 것이다. 전체적으로 조명의 조도를 낮추고, 일부 공간의 빛을 비워내 공간에 여유와 부드러운 색감과 따뜻함이 담길 수 있는 빛을 만들어 가기 시작했다.

전통적인 한국적 미학의 품격과 가치를 갖춘 공간
빛의 콘셉트는 아시아나항공이라는 기업이 추구하는 한국적 원형 경관의 아름다움을 담는 것이었다. 아시아나항공의 대표적 이미지인 색동저고리의 단아한 컬러(Color)에 담겨있는 한국적 미학의 품격과 가치를 빛으로 담아내는 것이었다. 빛은 단순히 비추는 대상과 보이는 관계가 아니라, 공간을 이야기하는 주체로써 빛은 공간 속에 존재한다. 그리고 사람들에게 핵심적인 가치를 느끼게 해주는 요소로, 빛은 어느 곳에서도 동일한 가치로 존재하지 않는다. 그래서 우리는 항공사 라운지의 빛을 공간마다 조금씩 다르게 디자인하여 각각의 공간에서 평온을 느끼도록 했다.

LIGHT PRODUCTION OF ASIANA AIRLINES LOUNGE AT INCHEON INTERNATIONAL AIRPORT

Create a Classy Space with Traditional Korean Aesthetics
With the relocation of Korean Air's lounge from Incheon International Airport's Terminal 1 to Terminal 2, Asiana Airlines' First Class, Business Class, and Star lounges have been created in the new space. Asiana's expectation was to create a space with a distinctly Asian flavor with traditional Korean aesthetics. Before designing Asiana's airline lighting, we analyzed both the existing interior lighting in the airline's lounges and the interior lighting of Incheon International Airport Corporation. The existing design was uniform and pale in color, with pale lighting. We realized that the lighting was installed simply as a lighting function without expressing the characteristics of the interior finishing materials.

Create a Warm and Welcoming Atmosphere for Your Customers
We wanted the light to reflect Asiana Airlines' 'heartfelt commitment to welcoming customers with care and sincerity', and to create a sense of empathy for those who are anxiously awaiting a long flight. By lowering the overall lighting level and emptying out the light in some spaces, we began to create a light that would allow the space to be relaxed, soft colors and warmth.

A Space with the Dignity and Value of Traditional Korean Aesthetics
The concept of light was to capture the beauty of the Korean prototype landscape that Asiana Airlines pursues as a company. The dignity and value of Korean aesthetics, contained in the simple colors of the Saegdongjeogoli (women's type of hanbok with colorful stripes by patch-working), the representative image of Asiana Airlines, were captured through light. Light is not just a visible relationship with the object it illuminates, but it exists in space as a subject that tells the story of space. And as an element that makes people feel core values, light is not the same everywhere. Therefore, we designed the light in Asiana Airlines lounges to be slightly different in each space to create a sense of tranquility in each space.

아시아나항공 에어버스 A330 | Asiana Airlines Airbus A330

아시아나항공의 첫인상을 빛 연출로 담아내다

나는 항공사의 대표적인 공간 라운지에 두 가지 빛 설계 원칙을 설정했다. 하나는 빛의 비움이고, 또 하나는 빛의 채움이다.

공간을 비우는 빛

제1터미널은 인천국제공항공사의 건축물이어서 안전상의 이유로 건축물 자체에 조명기구를 설치할 수 없었다. 전체 공간의 안전과 편의성을 만들어야 한다는 제약 때문에 천장에 조명을 설치하지 못하면서도 여러모로 어려움이 뒤따랐다. 이런 특수성 때문에 우리는 천장에 조명기구를 설치하지 않고 라운지의 빛을 만들어야 하는 비움의 방식으로 빛 디자인을 시작했다. 빛의 디자인은 현장에서의 다양한 마감 재질에 대한 빛 테스트를 통해서 공간 하나하나에 적절한 빛을 찾아내는 방법으로 빛 연출이 진행됐다. 게다가 국제공항이라는 특성 때문에 현장 빛 디자인 실험은 주로 이용객이 없는 새벽 시간에 이뤄져야 했다. 이런 여러 제약과 장소의 특수성 때문에 전혀 반대의 개념인 비움이라는 관점에서 접근했다. 공간이 비워짐으로써 더 매력적이고 더 아름다운 공간이 탄생할 수 있다고 확신한 것이다. 그래서 기존 인천국제공항 제1터미널의 큰 창문과 새롭게 설치되는 인테리어 구조물 자체를 빛으로 만들어 보고 싶었고, 이러한 구상이 새로운 빛을 만들어 내는 작업의 시작점이 되었다.

마음을 담는 빛

아시아나항공의 대표적 브랜드는 한국의 전통적인 색동의 문양이다. 이 그래픽 모티브를 중심으로 아시아나항공의 가치와 품격을 담아내고 싶었다. 아시아나항공의 항공사 라운지 빛이 아시아나항공 본연의 가치, 한국의 전통적인 품격을 담아내는 공간이 되도록 빛의 연출과 색동의 색감을 대표적인 이미지로 빛을 디자인했다.

구조물이나 공간에는 더러 보이지 않는 빛이 숨어 있다. 빛 자체가 벽이 되고 작은 빛이 잠깐의 휴식에 도움을 줄 수 있도록 했다. 최대한 눈부심이 없도록 세심하게 빛을 다듬어 나갔다. 공간의 특성이 다른 곳은 빛의 밝기를 달리하여 차별화했고, 다이닝룸(Dining Room)과 같은 곳에서는 편안하게 식사와 음료를 이용할 수 있도록 했다. 다이닝룸의 빛은 음식의 미감과 함께 다양한 채소와 와인의 색감이 선명하게 드러나도록 연색성*이 높은 빛을 공간에 담아 시감(視感)이 충만한 빛을 만들어 냈다.

*연색성: 물체에 빛이 비추어져 시각적으로 색감을 인지하는 정도를 나타내는 단어이다. 연색성이 높다거나 좋다는 말은 인공조명이 태양의 빛처럼 다양한 빛의 스펙트럼을 가지고, 물체 본연의 색감을 잘 표현할 때를 나타낼 때 사용한다.

Creating a First Impression with Light for Asiana Airlines

I set two light design principles for the airline's signature space lounge. One is the emptying of light, and the other is the filling of light.

The Light that Empties the Space

Terminal 1 was built by Incheon International Airport Corporation, so we were not allowed to install luminaires in the building itself for safety reasons. The constraints of creating safety and comfort for the entire space presented a number of challenges, including the inability to install lighting on the ceiling. Because of this particularity, we started the light design with a blank slate, where we had to create light in the lounge without installing luminaires on the ceiling. The light design was done on-site through light testing on different finishes to find the right light for each space. Furthermore, due to the nature of the international airport, the on-site light design experiments had to be conducted in the early morning hours, when there were no passengers. The limitations in this particular location led us to approach the project with a different perspective on emptiness and its significance. We were convinced that by emptying the space, a more attractive and beautiful space could be created. Therefore, we wanted to make light out of the large windows of the existing Incheon International Airport Terminal 1 and the new interior structure itself, and this concept became the starting point for creating new light.

The light that Fills the Heart

Asiana Airlines' representative brand is the traditional Korean colorful pattern. We wanted to capture the values and dignity of Asiana Airlines around this graphic motif, so we designed the light in Asiana Airlines' lounge as a space that captures the values of Asiana Airlines and traditional Korean dignity.

Light is often hidden in structures and spaces. The light itself becomes a wall, and a small light can help you relax for a moment. We carefully refined the light to make it as glare-free as possible. Different light levels are used to differentiate different areas of the space, and in areas such as the dining room, we wanted to make it comfortable to eat and drink. The light in the dining room is highly color rendering* to bring out the colors of the various vegetables and wines along with the aesthetics of the food, creating a light that is full of visual sensations.

*Color rendering: The degree to which we visually perceive color when light shines on an object. High or good color rendering is used to describe when an artificial light has a wide spectrum of light, just like the sun's light, and reproduces the natural colors of an object well.

인천국제공항 아시아나항공 오픈 라운지 | Asiana Airlines Open Lounge at Incheon International Airport

라운지 빛의 눈부심 테스트 | Glare test of lounge lighting

고객이 느끼는 아늑한 공간

인천국제공항 제1청사에서 출국심사를 마치고, 2층에서 에스컬레이터를 타고 3층 아시아나 항공사 라운지로 오르면서 아시아나항공 로고 윙(Wing)의 부드러운 빛의 흐름을 타고 비상하는 모습을 볼 수 있다. 이것이 아시아나항공의 첫인상이다. 처음 맞이하는 아시아나 항공사 로비와 리셉션 공간은 항공사 라운지 이미지를 통해 항공사의 가치를 보여주는 공간이다. 고객 입장에서는 처음 대하는 공간이니만큼 따뜻한 공간이라야 한다. 그래서 빛은 전체적으로 부드러운 색감의 간접적인 빛을 사용하여 아늑한 공간으로 연출해 냈다.

컴퓨터 시뮬레이션과 실물모형에 빛을 투사하는 다양한 실험

인테리어 구조물과 빛을 하나로 만들어 가는 방법과 구조물 내부에만 빛이 머물게 하는 방법으로 새롭게 디자인했다. 아시아나 항공사와 관련된 사람들과의 지속적인 소통, 때로는 설득을 통해서 아시아나항공만의 특별한 빛을 탐색해 나갔다. 이런 빛을 실행하기 위해서 빛의 시뮬레이션 과정이 있었다. 컴퓨터 프로그램을 통한 조도 시뮬레이션과 함께 모형에 빛을 투사하는 다양한 실험을 통해 직접 빛을 보고 느끼며 빛 디자인을 진행한 것이다. 위와 같은 다양한 테스트는 빛 연출에 있어서 매우 중요한 과정이다.

현장에 나타난 그림자 얼룩

현장에서 진행된 마지막 빛 연출 테스트에서 우리가 디자인한 빛에 의해 만들어진 불필요한 빛의 그림자를 제거해야 하는 심각한 문제가 나타났다. 공간을 연출하기 위해 계획된 빛이 실험 중에는 한 번도 나타나지 않던 그림자 얼룩이 인테리어 구조물에 발생한 것이다. 이 그림자를 어떻게 처리할 것인지에 대해서는 모형 제작을 통해서 실제 빛과 그림자를 볼 수 있었다. 우리는 현장에서 다시 고민을 시작했고, 그 속에서 빛을 빛으로 지우는 방법을 찾아냈다. 그림자를 지우기 위해서 또 하나의 빛으로 구조물의 특징을 표현하고 필요하지 않은 그림자를 지우고 나서야 첫인상을 만드는 리셉션 공간이 아늑하게 조성됐다.

A Cozy Space for Your Customers

After clearing immigration at Incheon International Airport Terminal 1, I take the escalator from the second floor to the Asiana Airlines lounge, on the third floor, and see the Asiana Airlines logo, Wing, floating in a gentle stream of light. This is the first impression of Asiana Airlines. The lobby and reception area of the Asiana Airlines lounge is the first space that shows the value of the airline through the image of an airline lounge. It is the first time a customer sees it, so it should be a warm space. Therefore, indirect light with soft colors was used throughout to create a cozy space.

Various Experiments with Projecting Light onto Computer Simulations and Physical Models

We redesigned the internal structure to integrate light within. Through continuous communication and persuasion with people involved in Asiana Airlines, we explored the special light of Asiana Airlines. To realize this light, there were various light simulations. In addition to simulating light levels through computer programs, we conducted various experiments by projecting light onto a model to see and feel the light. These various tests are an important part of the light design process.

Shadow Stains in the Field

The final light production test on site revealed a serious problem: unnecessary light shadows created by the light we designed had to be removed. The light planned for the space created shadow spots on the interior structure that never appeared during the experiments. As for how to deal with these shadows, the mockup allowed us to see the actual light and shadows. We started thinking again on site and found a way to erase the shadows with light, using another light to express the features of the structure and erase the unnecessary shadows, creating a cozy reception area that creates a first impression.

아시아나항공 로비 윙 구조물 | Wing Structure at the lobby of Asiana Airline

로비 윙 구조물 빛의 모형 제작 | Model creation of the lighting for the lobby wing structure

빛의 연출을 위한 스터디 | Study for light production

현장 조명설치를 위한 최종 테스트 | Final test for on-site lighting installation

아시아나 비즈니스 클래스 라운지 | Asiana Business Class Lounge

정적인 한국의 미, 격자 문양

퍼스트 라운지에 들어서면 전통 한옥의 격자 문양의 창문이 나열되어 벽을 이룬다. 이곳의 복도와 VIP룸은 조선시대의 정궁인 창덕궁에서 가장 아름답고 평온한 공간인 낙선재(樂善齋) 공간을 모티브로 빛을 디자인해 냈다. 밝은 보름달 빛의 색감을 콘셉트로 빛이 격자형 창문에 스며드는 이미지를 빛으로 연출한 것이다. 한국을 방문하는 외국인과 오랜 시간 비행을 준비하기 위해 잠시 머무는 사람들에게 가장 정적이며 한국적인 공간인 궁궐, 그 안에다 절제된 미학을 담는 공간으로서의 낙선재의 평온함, 따뜻함, 절제미를 아시아나항공사 라운지에 담았다.

Static Korean Aesthetic, Grid Pattern

Upon entering the First Lounge, a wall of windows in the lattice pattern of a traditional Hanok forms a wall. The corridors and VIP rooms are lighted in a way that is inspired by the most beautiful and tranquil space in Changdeokgung Palace, the Joseon Dynasty's main palace. The color of the bright full moon was used to create the image of light permeating through the lattice windows. The tranquility, warmth, and understated aesthetics of the palace, which is the most static and Korean space for foreigners visiting Korea and those staying for a short time to prepare for a long flight, are captured in the Asiana Airlines lounge.

평온한 VIP 라운지 | Tranquil VIP Lounge

빛으로 공간의 확장감이 연출된 비즈니스 클래스 라운지 | Business class lounge where the sense of space is enhanced through light

한옥 격자 문양의 벽을 이용한 빛의 진입로 | Light entrance using the grid pattern wall of Hanok

연색성을 향상시켜 음식 본연의 색감을 연출한 다이닝 룸 | Dining room enhances color to showcase the natural colors of the food

빛의 조도를 낮춘 휴식공간 | Relaxing room with light that reduces illuminance

눈부심이 없는 라운지 홀 | Glare-free lounge hall

테이블 집중감을 연출한 빛의 테스트 | Light test to enhance focus on the table

라운지 공간에 흐르는 감성의 빛

하루의 햇빛은 일출에서 정오를 정점으로 일몰의 석양빛까지 빛의 밝기와 색감이 다양하게 변화한다. 아시아나항공사 라운지의 빛도 시간에 따라 변한다. 빛의 밝기는 공간에 따라 조금씩 밝아진다. 창문에 가까이 갈수록 조금씩 밝기가 변하며 전체 라운지와 조화를 이루도록 빛을 디자인했다. 가구와 벽체는 인테리어 설계에서 공간을 구분하는 주요한 요소이자 빛을 굴절시키고 흡수시키는 또 하나의 빛이 되기도 한다. 그래서 빛은 마감재와의 관계성이 중요하다. 재료의 마감재에 따라 빛이 반사되고 산란하면서 독특한 공간을 만들기 때문이다.

     우리는 기존의 인테리어 마감재 특성을 고려하여 빛의 굴절과 확산, 투과 테스트를 진행했다. 또한 빛과 함께 만들어지는 구조물에 의해 발생하는 일부의 그림자를 없애거나 최소화하기 위해 빛의 설치 위치와 각도를 고려한 빛의 디자인을 통해서 아시아나 항공사의 라운지 공간에 새로운 감성의 빛을 담았다.

정적인 공간이 빚어내는 특유의 가치와 품격

고즈넉한 창덕궁 후원의 평온함을 담고 있는 빛의 색감과 빛의 레벨은 또 다른 관점에서 우리가 담아내고자 했던 아시아나항공 특유의 가치와 품격이었다.

     우리는 빛을 연출하지만 빛 자체는 나타내지 않았다. 단지 공간이 드러나고 그 속에서 많은 사람이 아시아나항공의 따뜻한 마음을 느낄 수 있는 공간으로 만들고 싶었다. 이를 위해 우리는 고객의 자리에서 빛을 보고 느끼면서 빛을 디자인했다. 고객이 가장 편하게 쉴 수 있는 정적인 빛의 모습을 담은 릴렉스(Relax)한 공간, 음식의 시감(視感)을 고려하여 연색성을 선택한 다이닝 룸(Dining Room)의 색감을 느낄 수 있는 빛, 매거진 셜프(Magazine Shelf)에서 벽면을 이용한 라이트 월(Light Wall), 모든 곳에 빛이 존재하지만 빛이 없는 빛의 디자인이다. 우리는 이 공간에 우리가 꿈꾸는 빛을 함께 만들고자 했고, 그 꿈을 이곳에 담았다.

Emotional Light in the Lounge Area

Sunlight throughout the day varies in brightness and color from sunrise to peak to the setting sun. The light in Asiana Airlines lounges also changes over time. The brightness of the light gradually increases depending on the space. The light is designed to harmonize with the rest of the lounge, changing in brightness as you get closer to the windows. Furniture and walls are the main elements that define spaces in interior design, but they also refract and absorb light. This is why light's relationship with finishes is so important. Different finishes of materials reflect and scatter light, creating unique spaces.

     We conducted tests on the refraction, diffusion, and transmission of light by considering the characteristics of existing interior finishes. We also considered the installation location and angle of the light to eliminate or minimize some of the shadows caused by the structures that are created with the light, and through the design of the light, we brought a new sense of light to the lounge space of Asiana Airlines.

Unique Value and Quality Created by Static Space

The color and level of light that captures the serenity of the tranquil Changdeokgung Palace Back Garden is another aspect of Asiana Airlines' unique values and dignity that we wanted to capture.

     We created light, but not the light itself. We wanted to create a space where people could feel the warmth of Asiana Airlines through the light, but not the light itself. To achieve this, we designed the light by seeing and feeling the light from the customer's seat, and customer's place: a relaxing space with static light where the customer can relax most comfortably; a light that can feel the color of the dining room, which chose color rendering in consideration of the visual sensation of food; a light wall using the magazine shelf; and a light design where light is everywhere, but there is no light. We wanted to co-create the light of our dreams in this space, and this is where we put it.

인천국제공항 제1여객터미널 아시아나항공 리커 바와 비즈니스 라운지 | Liquor Bar and Business Lounge at Asiana Airlines, Incheon International Airport Terminal 1

인천국제공항 제1여객터미널 아시아나항공 오픈 라운지 | Open Lounge, Asiana Airlines, Incheon International Airport Terminal 1

시간을 넘어서다:
유네스코 세계유산 궁궐과 건축물

BEYOND TIME:
UNESCO WORLD HERITAGE PALACES AND ARCHITECTURE

경복궁, 궁중문화축전의 빛을 밝히다

GYEONGBOKGUNG PALACE, ILLUMINATING THE PALACE CULTURAL FESTIVAL

창덕궁 달빛기행, 빛으로 꽃피우다

CHANGDEOKGUNG PALACE, WALKING IN THE ENCHANTING BEAUTY OF MOONLIGHT

운현궁, 흥선대원군의 이상(理想)을 빛으로 채우다

UNHYEONGUNG PALACE, FILLING THE IDEALS OF HEUNGSEON DAEWONGUN WITH LIGHT

절제된 빛의 미학
유네스코 세계유산 궁궐과 건축물
빛의 동중정(動中靜)

The understated aesthetics of light:
UNESCO World Heritage Palaces and Buildings
Dongzhongzheng of Light

서울, 정도 600년 역사와 국가유산의 도시

국가유산에 대한 빛 연출은 문화유산에 생명을 불어넣는 일이다. 국가 문화유산에 단순히 안전과 기능이라는 두 관점에서 빛을 연출하는 것이 아니라 오랜 시간을 함께한 국가 문화유산 본연의 가치를 드러낸다는 관점에서 빛이 연출되고, 빛이 역사 문화적 가치를 담아내야 한다. 국가유산은 일반 건축물에 비해 섬세하고 심도 있는 빛의 계획이 수반되며, 숙고 끝에 디자인되어야 한다. 왜냐하면 국가유산은 한번 훼손되면 다시 복원될 수 없는 고유의 가치를 지니고 있기 때문이다.

국가유산에 대한 빛 연출에서 중요한 부분은 국가유산 자체가 지닌 역사와 상징성 그리고 완전성(Integrity)에 대한 모습을 최소한의 빛으로 빚어내는 것이다. 상징성과 역사성은 국가유산을 테마파크처럼 조명색상의 극명한 차이를 바탕으로 드라마틱(Dramatic)한 이미지를 연출하는 것이 아니라 오랜 역사와

Seoul, about 600 Years Old and a City of National Heritage

Lighting national heritage sites is about bringing cultural heritage to life. It is not just about safety and function, but also about revealing the intrinsic value of the national heritage that has stood the test of time, and the light should reflect the historical and cultural values. Compared to ordinary buildings, national cultural heritage sites require more delicate and in-depth light planning, and must be designed with careful consideration, because they have unique values that cannot be restored once damaged.

An important part of lighting a national heritage site is to minimize the amount of light that is used to reflect the history, symbolism, and integrity of the site itself. Symbolism and history are not about creating dramatic

흔적, 그리고 깊이 간직하고 있는 본연의 가치를 어둠 속에서 찾아 빛으로 생명을 불어넣어 주는 것을 말한다.

실제로 국가 문화유산 빛 연출 디자인 사례를 통해서, 우리가 무엇을 고민했는지, 그리고 어떻게 빛을 디자인했는지 실제 빛 연출한 작품을 통해서 빛의 이야기를 들려주고자 한다. 국가 문화유산 본연의 가치를 나타내기 위해서 우리는 빛을 새롭게 더하고 만들어 가려는 탐구의 자세로, 최소한의 빛 연출 방법으로 빛을 디자인하고자 했다.

images based on stark differences in lighting colors like a theme park, but about finding the long history, traces, and deeply held intrinsic values in the darkness and bringing them to life with light.

We would like to tell the story of light through the actual light production design of national cultural heritage sites, explaining what we thought about and how we designed the light. In order to represent the inherent value of national cultural heritage, we tried to design light with minimal light production methods, with an attitude of exploration to add and create new light.

빛의 디자인, 보물 제1호 서울시 흥인지문(興仁之門) | Light Design, Treasure No. 1: Heunginjimun Gate, Seoul, Korea, 2014

경복궁,
궁중문화축전의 빛을 밝히다

경복궁 전각의 부드러운 선형을 빛으로 물들이다

조선시대 수도 한양(漢陽)은 한국의 수도 서울의 옛 이름이다. 서울에는 경복궁(景福宮), 창경궁(昌慶宮), 창덕궁(昌德宮), 덕수궁(德壽宮)과 경희궁(慶熙宮) 5대 궁궐이 있다. 경복궁은 조선시대 정궁으로 왕이 머물던 중심이 되는 궁궐이다. 궁궐 본연의 품격과 궁궐 건축의 특징이 잘 지켜진 건물로, 한국을 방문하는 관광객이 많이 찾는 명소이기도 하다. 경복궁 내부 2층 경회루에서 경복궁의 모습을 바라보면 지붕과 처마의 부드러운 곡선이 첩첩이 중첩되어 조망된다. 먼 산을 바라볼 때 산의 능선이 서로 겹쳐지면서 부드러운 곡선으로 보이는 것과 비슷하다. 우리의 궁궐 빛 연출은 인위적이지 않은 자연스러운 빛으로 품격을 지켜주고, 그 본연의 숭고한 가치를 드러내는 감성의 빛 디자인이었다.

GYEONGBOKGUNG PALACE, ILLUMINATING THE PALACE CULTURE FESTIVAL

Light Bathes the Soft Linear Lines of Gyeongbokgung Palace's Major Buildings in Color

Hanyang, the capital of the Joseon Dynasty, is the old name of Seoul, the capital of South Korea. There are five main palaces in Seoul: Gyeongbokgung Palace, Changgyeonggung Palace, Changdeokgung Palace, Deoksugung Palace, and Gyeongheegung Palace. Gyeongbokgung Palace was the official palace of the Joseon Dynasty and the centerpiece of the king's residence. It is a building that has preserved its original dignity and features of palace architecture, and is a popular attraction for tourists visiting Korea. If you look at Gyeongbokgung Palace from Gyeonghoe-ro on the second floor inside the palace, you can see the gentle curves of the roof and eaves with the overlapping of the chapels. It is similar to the way the ridges of a mountain appear as smooth curves as they overlap each other when looking at a distant mountain. Our palace lighting should be a sensitive light design that preserves the dignity of the palace with a natural light that is not artificial and reveals its inherent sublime value.

국보 제223호 경복궁(景福宮) 근정전(勤政殿) | Gyeongbokgung Palace's Geunjeongjeon Building, National Treasure No. 223

경복궁 본연의 모습은 자연스러운 선의 미학

조선시대 정궁인 경복궁은 경복궁의 가치를 어떻게 보여줄 것인가와 어떻게 느끼게 해줄 것인가에 대해 많이 고민했다. 낮에는 경복궁 안의 여러 전각이 유사하게 인지된다. 그러나 밤에는 빛과 어둠으로 경복궁의 주요 전각이 주변보다 밝게 빛나면서 우리가 담아내고자 했던 경복궁 전체 윤곽과 함께 경복궁의 숭고한 가치와 정연한 분위기를 담백하게 드러나도록 빛을 연출했다.

빛 자체는 공간의 주연(主演)이 아니다. 특히 경복궁에서 섬세한 빛 디자인은 경복궁 본연의 품격과 가치를 담아내기 위해 공간마다 조연(助演)으로 존재하도록 했다.

The Original Appearance of Gyeongbokgung Palace is Natural Line Aesthetics

Gyeongbokgung Palace, the official palace of the Joseon Dynasty, was carefully designed to showcase the value and feeling of the palace. During the day, the various halls in Gyeongbokgung Palace may appear similar, but at night, the play of light and darkness allows the main building to stand out, revealing the entire outline of Gyeongbokgung Palace. This helps capture the sublime values and orderly atmosphere of the palace.

Light itself is not the main element of a space, and in Gyeongbokgung Palace in particular, the delicate light design is meant to be a supporting character in each space to capture the dignity and value of Gyeongbokgung Palace.

경복궁 건춘문(建春門) 주변 탐방로 | Trails around Gyeongbokgung Palace's Gunchunmun Gate

밤을 배려하는 경복궁의 빛 연출

경복궁 본연의 가치를 담는 빛과 경복궁의 역사 문화자원의 훼손을 최소화하는 빛을 염두에 두고 빛은 디자인되었다. 이에 따라, 역사 문화유산의 위엄과 가치를 위해 연출하는 정서적인 빛과 빛이 존재하되 빛 자체가 보이지 않거나 인지되지 않도록 최대한 빛을 숨기는 방법으로 빛을 디자인했다.

지금까지 진행된 국가 문화유산에 대한 야간조명의 대부분은 시각적으로 인지하기 전에 눈부신 조명 때문에 본연의 모습을 인지하지 못하도록 방해하는 조명의 문제를 안고 있었다. 또한 일부 업라이트 조명이 처마 부분에 과도하게 비추고 있어서 용마루와 지붕을 이루는 각각의 부재(部材) 등이 밤에는 보이지 않고 사라지는 현상까지 일어났다. 밤을 배려하는 빛은 국가 문화유산 경관의 정체성과 완전성이 훼손되지 않는 최소한의 빛으로 전체의 이미지를 나타내는 것이 중요하다.

경복궁 탐방로의 작은 빛의 길

관람객이 궁궐에 들어서서 경복궁의 전각까지 거닐 수 있는 길은 좀 어둡더라도 보행에 지장을 주지 않을 적은 양의 빛만 필요했다. 우리는 궁궐 내부에 일반 공원에서 사용하는 4m 높이의 공원 조명등을 설치하지 않고 경복궁의 가치를 보존할 수 있도록 낮은 곳에 숨겨진 빛으로 디자인했다. 관람객이 궁궐 내부를 들어서서 전각까지 걸어가는 사람들 보행 동선을 따라 안전을 확보하는 빛이다. 주변의 어둠 속에서 빛 확산 면의 크기와 형태, 그리고 등의 설치 간격에 따라서 시각적으로 다르게 인지되도록 진입 공간을 만들었다. 동일한 빛의 양으로 공간을 밝힐 수는 있지만 확산하는 빛의 위치와 넓이, 등 배치 간격의 변화를 통해서 공간의 분위기를 조금씩 다르게 연출해 냈다. 우리는 궁궐 내부를 들어서면서 전각까지 걸어가는 탐방로의 낮은 인지책*에서 흙길이 은은하게 보이도록 빛을 만들었다. 낮에는 느끼지 못하는 흙의 따스함이 밤에는 빛에 의해 흙의 굴곡과 질감으로 현대 도시의 길 이미지를 떠나 옛길을 걷는 느낌이 들도록 한 것이다.

*인지책(人止柵): 박물관 또는 미술관 등의 전시관에서 관람자의 접근을 막는 장치

Light production with consideration of night at Gyeongbokgung Palace

The light was designed with the intention of creating a light that captures the value of Gyeongbokgung Palace and minimizing the damage to the historical and cultural resources of Gyeongbokgung Palace. Therefore, the light was designed in a way that hides the light as much as possible so that the light itself is not visible or perceived, although there is an emotional light and light that creates the majesty and value of the historical cultural heritage.

Most of the previous night lighting for national cultural heritage sites suffered from dazzling lights that prevented people from recognizing the true nature of the site before they could visually perceive it. In addition, some uplights overly illuminated the eaves, causing the eaves and individual members of the roof to become invisible and disappear during the night. It is important that night-conscious lighting presents an overall image with minimal light that does not compromise the identity and integrity of the national heritage.

Gyeongbokgung Palace's Little Light Path

The pathway for visitors to enter the palace and walk to the main hall of Gyeongbokgung Palace required only a small amount of light that would not impede walking in the dark. We did not install 4-meter-high park lights inside the palace that are used in regular parks, but instead designed low, hidden lights to preserve the value of Gyeongbokgung Palace. The light is meant to provide safety along the pedestrian path as visitors enter the palace and walk to the main hall. In the surrounding darkness, the entry space is visually perceived differently depending on the size and shape of the light diffusion surface and the spacing of the lights. The same amount of light can illuminate the space, but the location and width of the spreading light, as well as the spacing of the light fixtures, creates a slightly different atmosphere in the space. As we entered the palace, we created light to faintly illuminate the dirt path from an awareness barrier* on the walkway to the main hall. The warmth of the dirt, which is not felt during the day, is illuminated by the light at night, and the curves and textures of the dirt create the feeling of walking on an old road, leaving the image of a modern city street.

*Awareness barrier: A device that prevents visitors from entering an prohibited area in a palace to secure national cultural heritage and lawn.

경복궁 비현각(丕顯閣) 주변 탐방로 | Biheongak Surrounding Trail, Gyeongbokgung Palace

*법궁(法宮): 왕이 정규적으로 임어(臨御)하는 도성 내의 궁궐들 중에서 으뜸이 되는 궁궐.
*Beobgung: The head among the palaces.

경복궁 비현각(丕顯閣) 외부 공간, 담장의 빛

경복궁의 빛 디자인은 단순히 역사 문화경관 테마파크 속에 하나의 경관 요소처럼, 단순하게 조명하여 화려하게 연출하는 경관조명을 지양하고, 조선왕조의 법궁(法宮)*이라는 준엄한 가치를 담고 싶었다. 궁궐의 품격을 드러내고자 빛의 설치 위치를 처음부터 낮추도록 디자인했다. 우리가 바라보는 시각보다 조금은 아래 위치에 빛을 두고자 했다.

　　　　동정문을 통해 비현각의 외부 공간인 담장을 만나게 된다. 담장의 빛은 조금 다른 관점에서 별도로 생각했다. 바로 사람을 맞이하는 빛이다. 비현각 담장에 들어섰을 때 귀한 손님을 맞이할 때 걸어두는 청사초롱의 붉은빛과 푸른빛이 이 비현각 담장을 따라서 탐방객을 반갑게 맞이하는 이미지로 빛을 디자인했다. 비현각은 궁궐과 전혀 다른 빛의 접근으로 궁궐 내부와 외부 공간과의 차별성을 만들어 냈다.

Light on the Fence, outside of Gyeongbokgung Palace's Bihyeongak

The light design for Gyeongbokgung Palace is not simply a landscape element in a historical and cultural landscape theme park, and we wanted to avoid simple landscape lighting to create a flashy effect, but to capture the solemn value of the palace as a Beobgung* of the Joseon Dynasty. To showcase the dignity of the palace, we designed the light installation position to be low from the beginning. We wanted the light to be slightly lower than the viewer's perspective.

　　　　Through the Dongjeongmun Gate, we encounter the wall, which is an unrealized external space. The light of the wall was thought of separately from a slightly different perspective. It is the light that welcomes people. The red and blue colors of the Cheongsachorong lanterns, which are hung to welcome honored guests when they enter the Bihyeongak wall, were used to design the light along the wall to welcome visitors. The Bihyeongak has a completely different approach to light than the palace, creating a differentiation between the interior and exterior spaces of the palace.

경복궁 함화당(咸和堂)에서의 빛

경복궁 안 전각의 용도가 각기 다르듯 건축물의 특성도 각기 다르다. 그래서 공간의 특성에 맞춰서 공간의 이미지가 다르게 느끼도록 빛을 디자인했다. 함화당의 내부 공간은 LED 3,000K의 따뜻한 빛을 간접조명 방법으로 연출하여 부드러운 느낌이 들도록 했다. 진입부와 계단은 작은 조명기구를 제작하여 일부 시간만 빛을 연출하도록 했다. 함화당 실내 공간은 빛의 반사를 통해서 실내 공간이 은은하게 빛나도록 했다. 이는 공간 내부 마감이 밝은색으로 되어 있었기에 가능했다. 빛은 전체적으로 확산이 되는 낮은 밝기의 따뜻한 빛이다. 조명기구의 형태와 설치 위치를 변화시키는 현장 테스트를 통해서 함화당 내부에 각각 다른 빛이 놓이게 했다.

Light from Hamhwadang at Gyeongbokgung Palace

Just as each hall in Gyeongbokgung Palace has a different purpose, so do the characteristics of the architecture. Therefore, we designed the light according to the characteristics of the space so that the image of the space feels different. The interior space of Hamhwadang is indirectly lit with warm LED 3,000K light to create a soft feeling. The entryway and staircase were designed with small luminaires to create light only part of the time. The interior space of the Hamhwadang was designed to reflect the light, giving the interior space a subtle glow. This was made possible by the light color of the interior finishes. The light is a low brightness, warm light that diffuses throughout. Field tests were conducted to vary the shape and placement of the luminaires to create different light in the Hamhwadang.

경복궁 함화당(咸和堂) | Hamhwadang, Gyeongbokgung Palace

경복궁 집경당(緝敬堂) | Jipgyeongdang, Gyeongbokgung Palace

경복궁 집경당(緝敬堂) 실내에서 흘러나오는 빛

집경당 내부 공간의 빛은 목조건축을 훼손시키지 않도록 건축물 부재 중 대들보 위에 배치해서 간접적인 빛으로 공간을 밝혀냈다. 궁궐 건축의 마감 색채인 단청의 빛감을 은은하게 보여주기 위해 색의 재현도가 높은 광원(CRI 95)*을 사용하여 본연의 색감을 표현했다. 빛의 연출 방법도 탐방객의 시각에는 누수되는 빛**이 인지되지 않도록 빛 차단막을 사용했다.

목재와 유사한 마감 색상으로 처리하여 낮에는 조명기구가 보이지 않도록 하나하나 현장에서 설치할 위치를 찾아내고, 현장의 빛 테스트를 통해서 역사문화경관을 훼손시키지 않는 범위에서 빛을 완성해 냈다.

*CRI: Color Rendering Index, 태양의 빛은 다양한 파장을 가지고 있어서 파란 사과는 파랗게, 붉은 사과는 붉게 인지된다. 그러나 인공조명의 광원은 태양 빛의 파장을 모두 나타낼 수 없다. 태양 빛과 비교해서 색감을 나타내는 정도를 연색성이라 한다. 연색지수인 CRI가 낮은 광원에서는 파란 사과가 갈색으로 보인다.
**누수 되는 빛(Spill Light): 빛의 연출 시 의도하지 않은 빛으로 새어 나오는 불필요한 빛

Light Leaking from the Interior of Gyeongbokgung Palace's Jipgyeongdang

The light in the interior space was indirectly illuminated by placing it on the beams of the building's members to avoid damaging the wooden structure. To subtly show the shine of Dancheong, the finishing color of the palace's architecture, we used a light source with a high color rendering index (CRI 95)* to express its true color. The light was also directed using a light barrier to prevent any spill light** from being perceived by visitors.

The luminaires were treated with a finish color similar to wood so that they would not be visible during the day, and the location of each luminaire was identified on site, and the light was tested in the field to ensure that it would not damage the historical and cultural landscape.

*CRI: Color Rendering Index, the sun's light has different wavelengths, so a blue apple is perceived as blue and a red apple as red. However, artificial light sources cannot represent all the wavelengths of sunlight. The degree to which they display colors compared to sunlight is called color rendering. A blue apple will look brown in a light source with a low color rendering index, or CRI.
**Spill Light: Unnecessary light that leaks into the light that is not intended when producing light.

경복궁 집경당(緝敬堂) | Jipgyeongdang, Gyeongbokgung Palace

경복궁 소주방(燒廚房)과 생물방(生物房)에서의 빛

생물방 안에 들어섰을 때 빛의 밝고 화사한 관점에서 빛을 생각하고 만들어 냈다. 이곳은 연회가 열리는 공간으로, 다과(茶菓)를 함께 할 수 있다. 경복궁에서 유일하게 음식을 맛볼 수 있는 곳이다. 생물방 실내에 연색성이 높은 광원을 업라이트로 연출해서 실내 공간 전체가 밝은 이미지를 가질 수 있도록 빛 디자인을 진행했다. 이곳에 다운라이트 조명시설을 설치했다면 광원이 먼저 시각에 들어왔을 것이며, 눈부심으로 인해 공간 속에 숨어 있는 궁궐 건축의 아름다움을 전혀 느낄 수 없었을 것이다. 관람객의 눈에 광원이 잡히지 않았기 때문에 소주방과 생물방 실내 공간의 아름다움을 볼 수 있게 된 것이다. 우리가 빛에 대한 관점을 찾으려고 노력할 때 눈에 드러나지 않은 국가 문화유산 본연의 가치가 드러난다는 사실을 알아야 한다.

Light in the Gyeongbokgung Palace's Sojubang and Saengmulbang

When I entered the Saengmulbang, I thought of and created it in terms of light, bright and colorful. This is the space where banquets are held and refreshments are served. It is the only place in Gyeongbokgung Palace where you can taste food. The light design was carried out to create a bright image of the entire interior space by uplighting the Saengmulbang with a highly color rendering light source. If a downlighting fixture had been installed here, the light source would have been the first thing to come into view, and the glare would have made it impossible to appreciate the beauty of the palace architecture hidden in the space. Without the light source in front of the visitor's eyes, the beauty of the interior spaces of the Sojubang and Saengmulbang can be seen. We should recognize that when we try to find a perspective on light, the inherent value of national cultural heritage that is not visible to the eye is revealed.

경복궁 소주방(燒廚房)과 생물방(生物房)의 빛 | The Light of Gyeongbokgung Palace's Sojubang and Saengmulbang

경복궁 소주방(燒廚房) 탐방로 | Trail of Gyeongbokgung Palace's Sojubang

창덕궁 달빛기행, 빛으로 꽃피우다

창덕궁의 후원(後園), 금원(禁苑)의 가치를 빛으로 빚어내다

한국을 방문하는 외국 관광객이 가장 선호하는 밤의 장소*가 우리의 역사와 문화가 담겨있는 서울의 5대 궁궐이다. 그중 밤에 개장하는 유네스코 세계문화유산**인 창덕궁(昌德宮)의 달빛기행을 으뜸으로 꼽는다.

      창덕궁은 북한산 왼쪽 봉우리인 응봉(鷹峯) 자락에 자리 잡은 조선의 궁궐로, 1405년(태종 5년) 경복궁의 이궁으로 동쪽에 이웃한 창경궁과 서로 다른 용도로 사용되었으나 하나의 궁궐 역할을 해서 조선시대에는 '동궐(東闕)'이라고 불렀다. 임진왜란(1592년) 때 모든 궁궐이 불타고 광해군 때 다시 짓는 과정에서, 고종의 아버지인 흥선대원군이 경복궁을 중건하기 전까지 창덕궁이 조선의 법궁(法宮)으로써 오랫동안 사용했던 조선시대 왕실의 대표적인 궁궐이었다.

      경복궁의 주요 건축물이 좌우대칭의 일직선상으로 왕의 권위를 상징한다면 창덕궁은 응봉 자락의 지형에 따라 건물을 배치하여 한국 궁궐 건축 중 비정형적인 조형미를 대표하고 있는 궁궐 양식이다. 더불어 창덕궁의 후원(後園)인 금원(禁苑)은 각각의 권역마다 정자, 연못과 괴석 등이 잘 어우러진 왕실의 정원 역할을 하고 있다. 창덕궁은 현재 남아있는 조선의 궁궐 중 원형이 잘 보존되어 한국적인 정서가 담긴 궁궐로 평가받아 1997년 유네스코 세계문화유산으로 등재된 곳이다

CHANGDEOKGUNG PALACE, WALKING IN THE ENCHANTING BEAUTY OF MOONLIGHT

Changdeokgung Palace's Patronage, Bringing the Value of Geumwon to Light

Seoul's five palaces, which are deeply intertwined in our history and culture, are the most popular nighttime places* for international visitors to Korea. Among them, the moonlight tour of Changdeokgung Palace, a UNESCO World Heritage Site** that is open at night, is at the top of the list.

      Changdeokgung Palace is a Joseon palace located at the foot of Eungbong, the left peak of Bukhansan Mountain. It was built in 1405 (Taejong's five-year reign) as a second palace of Gyeongbokgung Palace, and although it served different purposes from its neighbor to the east, Changgyeonggung Palace, it was called 'Donggwol' during the Joseon Dynasty because it served as one palace. Changdeokgung Palace was the representative palace of the Joseon Dynasty, serving as the royal court for many years until the Imjin War(1592), when all the palaces were burned down and rebuilt during the Gwanghaegun, and Gojong's father, Heungseon Daewongun, rebuilt Gyeongbokgung Palace.

      While Gyeongbokgung Palace's main buildings symbolize the king's authority with their symmetrical, straight lines, Changdeokgung Palace is an atypical example of Korean palace architecture, with buildings arranged according to the terrain at the foot of Eungbong. In addition, the palace's backyard, Geumwon, serves as a royal garden with pavilions, ponds, and monoliths in each area. Changdeokgung Palace was inscribed as a UNESCO World Heritage Site in 1997, as it is considered to be one of the best-preserved remaining Joseon palaces in the world with a distinctly Korean sentiment.

---

*매일경제 MBN 뉴스, 2018.07.31.
**창덕궁(昌德宮): 1997년 12월 유네스코 세계문화유산으로 등재

*Maeil Business News, MBN News, 2018.07.31.
**Changdeokgung Palace : Listed as a UNESCO World Heritage Site in December 1997.

유네스코 세계문화유산 창덕궁(昌德宮) 전경 | View of Changdeokgung Palace, a UNESCO World Heritage Site

보름달 아래 펼쳐지는 창덕궁 달빛기행

창덕궁에서 진행하는 '창덕궁 달빛기행'은 한 달에 한 번 보름달이 뜨는 날, 일반인이 궁궐의 밤을 관람할 수 있는 독특한 콘셉트의 행사로, 2010년에 시작됐다. 은은한 달빛 아래 창덕궁 후원을 거닐며 창덕궁과 조선왕조의 역사 이야기를 듣고, 특히 예술적인 재능이 뛰어났던 조선왕조의 왕세자 효명세자의 춘앵전(春鶯囀)을 관람할 수 있는 고품격의 문화행사이다. 해가 갈수록 인기가 좋아서 행사장 입장권 예매 시작과 동시에 매진된다.

A Moonlit Walk through Changdeokgung Palace under the Full Moon

Changdeokgung Palace's Moonlight Walk is a unique concept that allows the public to experience the palace at night on a full moon once a month, and was launched in 2010. It is a high-quality cultural event that allows visitors to stroll around the grounds of Changdeokgung Palace under the soft moonlight, hear stories about the history of Changdeokgung Palace and the Joseon Dynasty, and watch the Chunaengjeon of Crown Prince Hyomyeong, who was particularly talented in the arts. The event has become so popular that it sells out as soon as tickets go on sale.

창덕궁 관람정(觀纜亭)의 가을 풍경 | Autumn scenery from the Gwanramjeong Pavilion, Changdeokgung Palace

창덕궁 성정각의 홍매화 | Red Plum Blossoms in Sungjunggak, Changdeokgung Palace

창덕궁 주합루(宙合樓) | Juhaplu, Changdeokgung Palace

창덕궁의 고요한 달빛 기행

창덕궁 '달빛기행' 시작 4년째 되던 2013년, 전 국가유산청인 문화재청의 창덕궁관리소로부터 '창덕궁 달빛기행 빛의 디자인 설계의뢰'를 받았다. 나는 창덕궁에 절제된 빛을 연출하여 보름달 아래에 은은하게 피어나는 궁궐의 품격을 새롭게 창출하고 싶었다. 먼저, 우리는 창덕궁 빛의 디자인에 앞서서 창덕궁의 밤을 위해 무엇을 중심에 둘 것인가를 두고 논의했다. 조선시대 효명세자가 꿈꾼 정제(整齊)된 미학의 세계를 중심에 두기로 했고, 우리의 선택은 절제된 빛으로 단아한 달빛의 부드러움이었다. 창덕궁의 전각과 담장, 후원의 숲과 산책로, 수변공간에 빛을 조금씩 덜어내는 방법으로 창덕궁에 달빛을 담아가기 시작했다.

       기존 창덕궁 건조물 일부 구간에 설치되었던 조명시설은 창덕궁 본연의 진정성(Authenticity)을 드러내는 범위에서 과감하게 빛을 낮추거나 숨겼다. 밤에 빛의 리듬감을 연출하기 위해 일부 조명은 다시 배치해서 아늑한 공간으로 느낄 수 있도록 연출했다. 공간으로 진입할 때 서서히 어두워지고 서서히 밝아지도록 빛의 리듬감을 특징 있게 도입했다. 창덕궁을 비추어 주는 빛은 있으나 조명기구가 외부로 노출되지 않도록 빛을 계획하여 창덕궁 달빛기행에 참가한 탐방객이 효명세자가 추구한 '정제된 미학'을 창덕궁 전각과 후원의 빛 속에서 느낄 수 있도록 시각적 암순응 기법을 통해 연출했고, 창덕궁 스스로 부상(浮上) 하도록 빛을 만들어 냈다.

A Tranquil Moonlight Walk at Changdeokgung Palace

In 2013, four years into the Changdeokgung Palace Moonlight Tour, I was commissioned by the Changdeokgung Palace Management Office of the National Heritage Administration, the former cultural heritage agency, to design the Changdeokgung Palace Moonlight Tour. I wanted to create a restrained light for Changdeokgung Palace, creating a new sense of the palace's dignity that blooms softly under the full moon. Before designing the light, we discussed what would be the centerpiece for the nighttime at Changdeokgung Palace. We decided to focus on the world of refined aesthetics envisioned by Joseon Dynasty Crown Prince Hyomyeong, and our choice was the softness of moonlight with its understated glow. We began to capture the moonlight in Changdeokgung Palace by gradually releasing light into the palace's halls and walls, the surrounding forest, trails, and waterfront spaces.

       The existing light fixtures installed in some sections of the palace buildings were drastically reduced or hidden to the extent that the authenticity of the palace was revealed. To create a sense of rhythmic light at night, some of the lights were rearranged to make the space feel cozy. The rhythm of light was characterized by gradual darkening and gradual brightening as you enter the space. Although there is light that illuminates Changdeokgung Palace, the lighting was planned in such a way that the luminaires were not exposed to the outside, so that visitors who participated in the Changdeokgung Palace Moonlight Tour could feel the "refined aesthetics" pursued by Crown Prince Hyomyung in the light of the Changdeokgung Palace's front and back halls, and the light was created to float on its own.

창덕궁 희정당(熙政堂) | Heejeongdang Hall, Changdeokgung Palace, 2013

창덕궁의 평온한 밤을 거닐다

효명세자의 정제된 미학을 느끼게 하기 위해서는 기존 국가 문화유산의 보여주기식 조명을 배제하고, 역사 속의 창덕궁과 말 걸기 하는 감성을 빛의 디자인으로 만들어 갔다. 이러한 감성을 연출하기 위해서 우리는 어둠 속에서 빛의 밝기를 서서히 어둠 속에 순응하도록 빛이 비추어지는 부분과 빛의 수량을 낮추는 방법으로 빛이 창경궁 속에 스며들어 궁궐 자체가 인지되도록 하는 연출을 만들어 냈다.

오백 년 전, 화사한 보름달 달빛 아래 조선왕조의 궁궐인 창덕궁을 효명세자와 함께 걷는 느낌을 주고자 했다. 밝고 화사한 공간에서 부드럽고 아늑함을 느끼는 밤의 후원으로 천천히 이동하며 시각적인 순응을 통해 우리의 시각은 편안해지고 숨어 있던 궁궐의 모습과 자연이 우리 마음에 들어오기 시작한다.

달빛에 젖은 풍광과 효명세자의 정제된 미학

창덕궁 밤의 풍광 콘셉트는 달(月)이다. 월태화용(月態花容)*에서 착안하여 은은한 이미지로 창덕궁의 아름다운 밤을 창출했다. 우리는 "창덕궁 달빛기행을 통해서 달빛 아래 피어나는 효명세자의 정제된 미학을 느낄 수 있어야 한다."라는 생각으로 빛을 만들어 갔다.

Stroll through the Tranquil Night of Changdeokgung Palace

In order to create a sense of the refined aesthetics of Crown Prince Hyomyung, we excluded the superficial illumination of the existing cultural heritage and created a light design that speaks to the historical Changdeokgung Palace. To create this emotion, we gradually adjusted the brightness of the light in the darkness to acclimate to the darkness by reducing the area and quantity of light, creating a production that allows the light to penetrate into the Changdeokgung Palace and recognize the palace itself.

We wanted to create the feeling of walking with Crown Prince Hyomyeong through Changdeokgung Palace, a palace of the Joseon Dynasty, under the light of a brilliant full moon, five hundred years ago. Through visual acclimatization, our eyes are relaxed and the hidden palace and nature begin to enter our minds as we slowly move from the bright and colorful space to the soft and cozy support of the night.

Moonlit Landscapes and the Refined Aesthetic of Crown Prince Hyomyung

The concept of the Changdeokgung Palace night landscape is the moon. Inspired by Woltaehwaryong*, we created a beautiful night scene of Changdeokgung Palace with a subtle image. We created the light with the idea that "people should be able to feel the refined aesthetics of the Crown Prince Hyomyung blooming under the moonlight through the moonlight walk in Changdeokgung Palace."

*월태화용(月態花容): 달 같은 태도(態度)와 꽃 같은 얼굴이라는 뜻으로, 달빛 아래 미인을 뜻함

*Woltaehwaryong : Meaning "moon-like attitude" and "flower-like face," referring to a beauty in the moonlight

창경궁의 정문, 돈화문(敦化門)에 들어서다

창덕궁에서 처음 반기는 곳이 창덕궁 정문인 보물 제383호 돈화문(敦化門)이다. 돈화문의 건축물 부분인 공포(栱包)에서 빛나는 단청(丹靑) 빛은 조선의 정궁을 나타내는 이미지이다.

따뜻한 색감과 밝은 빛의 이미지로 창덕궁 진입경관을 표현했다. 기존에 설치되었던 조명기구에서 발생하는 과잉된 빛 공해를 차단하기 위해 빛 차단막을 설치하고, 빛의 각도와 밝기를 낮추었다.

돈화문에서 금천교(錦川橋)에 이르는 일대는 왕실의 사무기관인 궐내각사가 모여 있는 외조(外朝)에 해당하는 곳이다. 이곳에 300년이 넘은 회화나무* 세 그루가 서 있다. 회화나무의 저층부와 상층부로 구분하여 회화나무의 자연스러운 수형의 아름다운 자태를 드러내 주기 위해 빛을 추가로 설치하고, 계절별로 나뭇잎의 형태가 변화하는 특성을 고려하여 빛을 디자인해 나갔다. 조명기구가 설치된 지점에서 반경 5m 내에서 자유롭게 이동이 가능하도록 조명기구를 별도로 디자인하여 설치했고, 개별적으로 빛의 밝기를 콘트라스트로 조절할 수 있도록 제작했다. 이런 특수 빛 연출을 통해 조망 거리에 따른 회화나무 수형의 입체감이 드러나도록 했다.

Entering Donhwamun Gate, the main entrance to Changdeokgung Palace

The first thing that greets you at Changdeokgung Palace is Donhwamun Gate, Treasure No. 383, the main gate of Changdeokgung Palace. The Dancheong light shining from the pillar head, the architectural part of Donhwamun Gate, is an image that represents the Joseon Dynasty's main palace.

The warm colors and bright light are used to represent the entrance to Changdeokgung Palace. To block excessive light pollution from the existing lighting fixtures, a light barrier was installed, and the angle and brightness of the light was reduced.

The area from Donhwamun Gate to Geumcheongyo Bridge is the outer court, where the royal government offices are located. There are three Sophora japonicas* that are over 300 years old. Additional lighting was installed to showcase the natural beauty of the trees' shapes by dividing them into lower and upper tiers, and the lighting was designed to take into account the changing shape of the leaves throughout the seasons. The lighting fixtures were designed and installed separately so that they could be moved freely within a radius of 5 meters from the point where they were installed, and the brightness of the light could be individually adjusted with contrast. Through this special light production, the three-dimensional sense of the tree's shape was revealed depending on the viewing distance.

*회화나무: 회화나무 열매는 학자가 밤에 불을 밝히는 기름으로 사용하기 때문에 일명 '학자나무'로 불리며, 명문가에서 훌륭한 학자 배출을 기원하여 심었다. 임금이 신하에게 내리는 하사목(下賜木)이 되었다.

*Sophora japonica: The fruit of the Sophora japonica tree is called "scholar's tree" because scholars use its oil to light their lamps at night, and it was planted by prestigious families in hopes of producing great scholars. It became a gift from the king to his subjects.

창덕궁 돈화문(敦化門) | Donhwamun Gate, Changdeokgung Palace

금천교(錦川橋)와 진선문(進善門) | Geumcheongyo Bridge and Jinsunmun Gate

금천교(錦川橋)에 드리워진 달빛

탐방객이 금천교(錦川橋)인 석교(石橋)를 지나간다. 금천교 위에 돌 형태로 만들어진 작은 빛을 바라본다. 빛은 금천교의 석수(石獸)와 석교 난간의 장식을 업라이트(Up Light)로 비추고 있어서 석교 돌 장식의 모습은 낮과 다른 신비감과 성스러운 공간으로 들어가는 느낌을 만들어 준다. 우리는 금천을 건너 오백 년 전 조선시대 속으로 들어선다. 진선문(進善門)이 눈에 들어오고, 진선문에서 멀리 숙장문(肅章門)이 보인다. 문이 연속으로 중첩된 경관과 함께 숙장문 현판을 비추는 낮은 조도의 빛 연출은 창덕궁을 찾는 사람들에게 달빛 아래 은은하게 빛나는 창덕궁의 아름다움을 느끼게 해준다.

　　　　이제 시간을 거슬러 과거로 향한다. 작은 빛이 서서히 어둠 속으로 스며들면서 주변의 조도가 함께 낮아진다. 우리는 어둠에 젖어 들었지만 아직 어둠을 인지하지 못한다. 우리 눈이 이미 어둠 속에 순응되어 있기 때문이다. 나는 시각적 순응 연출을 통해서 편안한 달빛 아래 창덕궁을 느낄 수 있도록 빛을 디자인했다. 이제, 창덕궁의 정전인 인정전(仁政殿)을 인정문을 통해 바라보게 된다. 인정전의 근엄한 빛은 행랑채 전각 뒤에 높이가 조절되는 조명기구에 의해 인정전의 상부 2층 누각이 빛나도록 연출을 해냈다. 변화하는 인정전의 빛을 통해 낮에는 볼 수 없었던 인정전 2층 누각의 경이로운 한국 궁궐 건축의 장엄하고 아름다운 단청과 용마루의 선형, 기와지붕 처마의 곡선미를 볼 수 있는 것이다.

　　　　우리는 이러한 연출을 위해 인정전 회랑에 설치되어 있던 기존 조명에 빛 차단막을 추가로 설치했다. 전체적으로 빛의 위계를 고려하여 인정전 외부에 설치된 조명도 일부 소등하고, 일부는 밝기를 조절하여 전체적으로 궁궐의 건축양식이 빛에 드러나도록 빛을 디자인해 냈다. 그리고 인정전 내부에도 기존의 전기선을 활용한 빛을 추가로 설치하여 그동안 볼 수 없었던 인정전 천장에 조각된 봉황의 조각을 감상할 수 있도록 만들었다.

　　　　빛의 색상은 궁궐의 붉은빛인 주칠의 빛과 화려하지만 단아한 색감을 지닌 단청색 본연의 색감을 나타낼 수 있도록 부드럽고 따뜻한 빛(3,000K)을 낮은 조도 레벨로 만들어 냈다.

Moonlight on the Geumcheongyo Bridge

A tourist walks across the stone bridge, the Geumcheongyo Bridge, and gazes at a small light made in the form of a stone on the bridge. The light illuminates the stonework of the bridge and the decorations on the stone railings, creating a sense of mystery and entering a sacred space that is different from the daytime. As we cross the Geumcheon Stream, we are transported back to the Joseon Dynasty, five hundred years ago. Jinsunmun Gate comes into view, and from Jinsunmun Gate we can see Sookjangmun Gate in the distance. The view of the gates stacked on top of each other and the low light illuminating the Sookjangmun Gate plaque gives visitors a sense of the beauty of Changdeokgung Palace, which shines softly under the moonlight.

　　　　Now, let's go back in time. As the small light slowly seeps into the darkness, the light level around us decreases. We are immersed in darkness, but we are not yet aware of it. Our eyes have already adapted to the darkness. I designed the light to create visual acclimatization so that we can feel Changdeokgung Palace under a comfortable moonlight. Now, we are looking at the main hall of Changdeokgung Palace, the Injeongjeon Hall, through the Injeongmun Gate. The solemn light of Injeongjeon wall is created by a height-adjustable luminaire behind it, allowing the upper two floors of the pavilion to shine. The changing light of Injeongjeon allows the viewer to see the majesty and beauty of Korean palace architecture, the linearity of the Dancheong and Yongmaru, and the curvature of the tiled roof eaves, which are not visible during the day.

　　　　To achieve this effect, we installed additional light shields to the existing lighting in the corridor of Injeongjeon Hall. Considering the overall hierarchy of the light, we also designed the lighting outside the palace by turning off some lights and adjusting the brightness of others so that the architecture of the palace is revealed in the light. Inside the palace, additional lights were installed using existing electrical lines to allow the viewer to see the phoenix carvings on the ceiling of the palace, which had never been seen before.

　　　　As for the color of the light, a soft, warm light (3,000K) was created at a low illumination level to represent the reddish glow of the palace's lacquer and the original color of the palace's brilliant but simple blue.

인정문(仁政門) | Injeongmun Gate

인정문과 회랑(回廊) | Injeongmun Gate and Corridor

인정전(仁政殿)에서 보는 서울의 아름다운 야경

인정전의 2층을 비추어 주는 빛은 우리가 바라보는 각도보다 높였기 때문에 조명기구의 광원이 탐방객에게는 보이지 않는다. 어둠에 순응된 탐방객의 시각에 평온함을 지켜줄 수 있도록 특별하게 빛을 디자인한 것이다. 이런 시각 순응의 특성을 고려한 절제된 빛 연출을 통해 관람객은 인정전에 올라서서 멀리 남산 N서울타워를 바라보게 된다. 그리고 멀리 보이는 서울 야경은 창덕궁의 풍광과 대비되어 더욱 차분하게, 고상한 공간이 되도록 했다.

Beautiful Night View of Seoul from Injeongjeon Hall

The light that illuminates the second floor of the Injeongjeon Hall is elevated above our viewing angle, so the source of the light fixture is not visible to visitors. The light is specially designed to protect the calmness of the visitor's eyes, which are acclimatized to darkness. Through the restrained light production that takes into account this characteristic of visual adaptation, the visitor can stand on the Injeongjeon hall and look at Namsan 'N Seoul Tower' in the distance. The distant night view of Seoul is contrasted with the scenery of Changdeokgung Palace to create a more calm and noble space.

완전성을 연출한 국보 제225호 창덕궁 인정전(仁政殿)
National Treasure No. 225, portrayed with completeness, Injeongjeon Hall, Changdeokgung Palace

달빛에 젖은 선정전(宣政殿)과 희정당(熙政堂)의 풍광

인정전을 돌아 선정문과 희정당(熙政堂)을 거닐면서 어두운 밤의 정경을 자연스럽게 바라본다. 이러한 풍광을 구현하기 위해 선정전(宣政殿)의 빛은 선정문 내부에 LED 조명 9W(3,000K)의 빛을 사용했다. 궁궐 주변은 조도가 낮기 때문에 빛의 대비(Contrast)를 낮추는 작은 빛으로 공간감을 나타낸 것이다. 달빛 젖은 눈에 선정전과 희정당의 아름다운 풍광이 더 고즈넉하다.

Moonlit View of the Seonjeongjeon Hall and Heejeongdang

As you walk around the Injeongjeon Hall and through the Seonjeongmun Gate and Heejeongdang, you will naturally see the scene at night. To realize this scenery, the light in the Seonjeongjeon Hall uses 9W(3,000K) of LED lights inside the gate. Since the area around the palace has low light levels, the small light, which lowers the contrast of the light, creates a sense of space. The beautiful scenery of the Seongjeongjeon Hall and Heejeongdang is more serene in the moonlight.

나무의 색감을 나타낸 선정문(宣政門) | Displaying the colors of the wood, Seonjeongmun Gate

창덕궁 낙선재(樂善齋)의 빛은 아늑하고 따뜻한 여성의 공간

창덕궁 달빛기행의 빛은 모두 창덕궁 스스로가 빛을 내는 구조로 빛을 디자인했다. 어둠 속에서 평온하게 시각의 순응이 이루어지고, 우리는 오백 년 전의 모습을 어둠 속에서 만나게 된다. 그리고 부드러운 창호지를 통해서 자연스럽게 새어 나오는 빛을 보게 된다. 바로 창덕궁 여성만의 공간이었던 낙선재(樂善齋)의 빛이다. 낙선재 실내 공간에 연출된 아늑하고 따스한 빛은 낙선재 창문과 창틀을 부드러운 빛으로 물들인다. 밤을 배경으로 창호지에서 배어 나오는 빛에 의해 은은하게 부상하는 전통 문양의 빛은 창덕궁만의 풍광이다.

Changdeokgung Palace Nakseonjae's Light is a Cozy and Warm Women's Space

The light of Changdeokgung Palace Moonlight Tour is designed in a way that the palace itself emits light. In the darkness, visual adaptation occurs peacefully, allowing us to encounter the appearance from five hundred years ago. We can see the light gently flowing through the soft window lattice. It is the light of Nakseonjae, a space exclusively for women in Changdeokgung Palace. The cozy and warm light created in the interior space of Nakseonjae softly illuminates the windows and window frames. The traditional patterns gently emerging with the light flowing from the latticed windows against the backdrop of the night are the scenery unique to Changdeokgung Palace.

빛에 의해 연출된 낙선재(樂善齋) 전통문양 | Traditional patterns of Nakseonjae Hall created by light

빛이 없어도 넉넉한 부용지(芙蓉池) 가는 길

낙선재에서 승화루(承華樓)를 거쳐 창덕궁의 후원 산책로를 따라 주합루(宙合樓)로 향하는 길은 창덕궁과 창경궁이 접한 길이다. 양쪽으로 담장이 있는 창덕궁 후원을 오르는 길이다. 조금은 어두운 공간인 이곳을 더 어둡게 빛을 디자인하고자 했다. 왜냐하면 이곳 아래로 부용지(芙蓉池)를 만나기 때문이다. 부용지 수면에 반사된 주합루의 모습을 더 극적으로 보여주기 위해서이다. 그래서 빛을 바꾸어 디자인했다.

부용지 가는 탐방로에 어디에도 빛이 없다. 단지 담장 너머에 설치된 조명기구가 큰 회화나무를 비출 뿐이다. 담장을 넘어온 빛이 나무의 잎과 가지에 붙어서 은은하게 빛난다. 빛의 반사(Reflection)를 통해서 탐방로를 밝히도록 했다. 우리가 주변 빛과의 상관관계를 검토하고, 실제 현장에서 빛 테스트를 통해서 주변이 어두운 상황에서도 나뭇잎을 적시고 떨어지는 간접적인 빛만으로도 보행의 안전을 충분히 확보할 수 있다는 사실을 확인했기 때문이다.

A light-free path to Buyongji Pond that can be reached by moonlight

The path from Nakseonjae to Seunghwaru then along the Changdeokgung Palace Huwon walkway to Juhapru is the path that goes between Changdeokgung Palace and Changgyeonggung Palace. The path climbs up the Changdeokgung Palace Huwon, which is flanked by walls on both sides. I wanted to design the light to be darker in this somewhat dark space, because below it is Buyongji Pond. I wanted to show the reflection of Juhapru on the water of Buyongji Pond more dramatically. So I designed the light differently.

There is no light anywhere along the trail to Buyongji Pond. The only light is from a luminaire on the other side of the fence, which illuminates the enormous Sophora japonica tree. The light from the fence catches the leaves and branches of the tree and gives it a soft glow. The reflection of the light illuminates the pathway. We examined the correlation with the ambient light and confirmed through light tests in the field that indirect light from leaves is sufficient to ensure the safety of walking even in dark conditions.

돌의 질감을 표현한 불로문(不老門)과 애련정(愛蓮亭) | Expressing the texture of stone, Bullomun Gate and Aeryeonjeong

빼어난 모습을 담아낸 승재정(勝在亭) | Seungjaejeong, capturing an outstanding view

창덕궁 애련정(愛蓮亭)의 정취 애련지(愛蓮池)에 담다

밤 속의 창덕궁 모습을 빛의 리듬감으로 느끼게 해주고 싶었다. 궁궐의 전각이 모여 있는 궐내각사(闕內各司)와 창덕궁 후원 탐방로와 부용지(芙蓉池)와 애련지(愛蓮池) 주변 수변공간에 대해서는 물에 반사되는 빛의 효과를 나타내도록 빛의 위치와 연출에 변화를 주어 정적인 공간으로 느끼도록 빛을 창출해 냈다.

Capturing the atmosphere of Aeryeonjeong, Changdeokgung Palace in the Aeryeonji Pond

We wanted to create a sense of rhythmic light in the Changdeokgung Palace at night, and for the waterfront spaces around Buyongji Pond and Aeryeonji Pond, where the palace's former Gueolnaegaksa are located, as well as the Changdeokgung Palace Huwon Trail, we created light in a way that makes them feel like static spaces by changing the positioning and direction of the light to show the effect of light reflecting off the water.

창덕궁 애련지(愛蓮池)와 애련정(愛蓮亭) | Aeryeonji Pond and Aeryeonjeong Pavilion, Changdeokgung Palace

부용지(芙蓉池) 물빛에 담긴 주합루(宙合樓)

조선왕조의 서고(書庫)였던 2층 누각인 주합루(宙合樓)는 부용지 수면에 어리비쳐 또 하나의 주합루를 만들어 낸다. 부용정과 어수문과 주합루에는 조명이 일부 설치되어 있었다. 인공조명의 광원이 보이는 곳과 주합루 건축 형태가 잘못 드러나는 곳의 빛을 개선하고, 관람객 시점에서 주합루의 건축물의 위용이 자연스럽게 드러나도록 주합루와 부용정 실내 공간에도 작은 빛을 설치했다.

Buyongji Pond and Juhapru, Changdeokgung Palace

A two-story building that was a library during the Joseon Dynasty, Juhapru, is located on the water surface of Buyongji Pond, creating another Juhapru. Buryongjeong, the Eosumun gate, and the main hall were partially illuminated. Small lights were also installed in the interior spaces of the Juhapru and Buryongjeong to improve the light in places where the source of artificial light was visible and the architectural form of the main buildings were incorrectly revealed, and to naturally reveal the majesty of the main buildings from the viewer's perspective.

창덕궁 부용지(芙蓉池)와 주합루(宙合樓) | Buyongji Pond and Juhabru Pavilion in Changdeokgung Palace

창덕궁 후원(後苑)의 빛

애련지(愛蓮池)를 지나면 효명세자가 독서하던 단층 목조건축물인 의두합(倚斗閣)이 나타난다. 단청도 하지 않은 단아한 의두합이라는 공간적 특성을 실내에서 창호지를 통해 새어 나오는 빛으로 연출하고, 앞에 보이는 애련정을 애련지의 물속에 빛을 뿌려 간접적으로 담아냈다.

연경당(演慶堂)으로 향하는 장락문(長樂門)을 통과하면, 연경당에서 효명세자 춘앵전(春鶯囀)의 정제 공연을 관람할 수 있다. 기품이 흐르는 효명세자의 풍류를 음미하며 후원 산책로를 따라 탐방객은 천천히 현재의 시간으로 돌아오게 된다.

달빛기행의 마지막 코스인 창덕궁 후원 길은 아름드리 나무 숲속으로 난 길이다. 우리는 길을 보여주는 것이 아니라 나무가 빛이 되어주기를 바랐다. 우리가 디자인한 빛은 창덕궁 후원의 사계절 자연 속 나무와 숲에 있다. 숲에서 흘러나온 빛이 탐방객의 보행을 안내한다. 이러한 빛 연출이 가능했던 것은 탐방객의 시각이 이제 달빛에 적응하여 어둠 속에서도 부드럽고 편안함을 느끼고 있었기 때문이다.

Light of Changdeokgung Palace Huwon Garden

After passing the Aeryeonji Pond, the single-story wooden building, Uiduhap where Crown Prince Hyomyung used to read, appears. The spatial characteristic of the simple, unadorned building was captured by the light leaking through the windows from the inside, and the Aeryeonjeong seen in front of it was indirectly captured by scattering light into the water of the Aeryeonji Pond.

If you pass through the Jangnakmun Gate to Yeongyeongdang, you can watch a performance of the Chunaengjeon of Crown Prince Hyomyeong in Yeongyeongdang. The Huwon walkway slowly brings visitors back to the present as they savor the graceful flavor of the court.

The last destination of the moonlight tour, the Changdeokgung Palace Huwon Trail, leads through a forest of beautiful trees. We wanted the trees to be the light, not the path. The light we designed is in the trees and forests of Changdeokgung Palace's four-season nature. The light from the forest guides the visitor's walk. This was possible because the visitor's eyes had now adapted to the moonlight and felt soft and comfortable in the dark.

승화루(承華樓)와 상량전(上凉亭) | Seunghwaru Pavilion and Sangryangjeon

창덕궁 후원(後苑) 산책로 | Trail in Changdeokgung Palace Rear Garden

창덕궁 본연의 자태를 빛으로 담는다

창덕궁 자체가 아름다운 빛이 되어야 한다는 생각으로, 눈부심을 유발하는 조명기구는 각도를 변화시키거나, 일부 조명기구는 설치 위치를 다시 조정하는 방법으로 개선했다. 노출이 심하고 과도한 빛을 뿜는 조명기구는 철거하여 창덕궁 본연의 아름다운 모습이 더 이상 훼손되지 않도록 했다. 그리고 최대한 창덕궁 내부에서 스며 나오는 빛을 통해 살아 숨 쉬는 궁의 모습을 담아냈다.

Capturing the Essence of Changdeokgung Palace through Light

The idea was that the palace itself should be a beautiful light, so fixtures that caused glare were angled differently, and some fixtures were repositioned. Exposed and excessive light-emitting fixtures were removed so that they would no longer detract from the palace's natural beauty, and as much as possible, light from within the palace was used to capture the palace as a living, breathing space.

보름달을 바라보는 상량정(上凉亭) | Viewing the full moon, Sangryangjeong

대금이 연주되는 영화당(暎花堂) | Yeonghwadang, where Daegeum is played

운현궁, 흥선대원군의 이상을 빛으로 채우다

운현궁(雲峴宮)은 조선시대의 정궁(正宮)이 아니다. 고종황제(1852~1919)의 아버지 흥선대원군의 사가(私家)로, 고종이 태어나서 왕위에 오를 때까지 성장한 공간으로, 사적 제257호로 지정되어 있다. 2012년부터 도심 속 아름다운 쉼터를 표방하며 한여름 밤 시민들에게 작은 공연과 음악회를 개최하여 국가 문화유산 관람의 문턱을 낮춘 공간이기도 하다.

과거 야간 개방 이전에는 운현궁의 아름다움을 제대로 표현할 수 있는 조명이 설치되지 않았다. 일부 조명시설은 운현궁 전체를 밝혀주지 못하고 한곳에만 집중하여 투사되거나 과잉된 빛이 주변으로 넘쳐나고 있었다. 시민들이 밤의 운현궁을 관람하기에는 여러모로 불편한 점이 많았다. 우리는 운현궁 주요 공간과 인물에 대한 자료 검토를 시작으로 빛 연출 계획을 수립해 나갔다. 빛의 콘셉트는 고종황제의 부친 흥선대원군 이하응(李昰應, 1821~1898)의 새로운 왕도정치에 대한 이상을 담는 것이었다.

흥선대원군의 품격과 가치, 노안당(老安堂)에 빛으로 담아내다

기존에 설치된 조명이 단순히 대상을 밝히는 것이었다면, 우리가 꿈꾼 빛은 조선 후기에 주요한 정책이 만들어진 역사적인 공간인 운현궁에 흥선대원군의 강인한 성품을 느낄 수 있는 담백한 빛을 창출하고 싶었다. 운현궁 정치의 중심 공간인 노안당(老安堂)에 새로운 빛을 담았다. 빛이 공간을 단순히 밝히는 것이 아니라, 빛과 주변 환경이 조화를 이룰 수 있는 하나의 격조 있는 분위기를 만들어 내고자 했다. 현장에서의 협의와 빛의 테스트, 그리고 디자인 컨설팅인 총감독을 통해 빛 연출을 위한 최적의 조명설치 장소와 빛의 밝기, 빛에 의해 표현되는 색감 탐색에 나섰다.

UNHYEONGUNG PALACE, FILLING THE IDEALS OF HEUNGSEON DAEWONGUN WITH LIGHT

Unhyeongung Palace is not the official palace of the Joseon Dynasty. It is the private residence of Emperor Gojong's (1852-1919) father, Heungseon Daewongun, where Gojong was born and grew up until his ascension to the throne, and is designated as Historic Site No. 257. Since 2012, it has been a beautiful shelter in the center of the city, offering small performances and concerts to the public on midsummer nights, lowering the threshold for viewing national cultural heritage.

Prior to its nighttime opening, Unhyeongung Palace was not properly lit to showcase its beauty. Some of the lighting fixtures were not able to illuminate the entire palace, instead focusing on a single area and projecting excessive light into the surrounding area. There were many inconveniences for the public to see the palace at night. We started by reviewing the main spaces and figures of Unhyeongung Palace and developed a lighting plan. The concept of the light was to capture the ideals of Emperor Gojong's father, Heungseon Daewongun, who envisioned a new monarchy.

The Dignity and Value of Heungseon Daewongun, Noandang, Captured with Light

Whereas the existing lighting simply illuminated the object, we wanted to create a light that conveyed the strong character of Heungseon Daewongun in Unhyeongung Palace, a historic space where major policies were made in the late Joseon Dynasty. We created a new light in the Noandang, the central political space of Unhyeongung Palace. We wanted to create a dignified atmosphere in which the light not only illuminates the space, but also creates a harmony between the light and the surrounding environment. Through on-site consultations, light tests, and design consulting, we explored the best place to install the light, the brightness of the light, and the colors expressed by the light.

조선의 역사를 품은 시간의 흔적, 운현궁 노안당(老安堂)
Traces of time that carry the history of the Joseon Dynasty, Noandang, Unhyeongung Palace

빛의 디자인, 담백한 빛으로 담아낸 운현궁(雲峴宮) 노안당(老安堂) | Light Design, Noandang captured in simple and refined light, Unhyeongung Palace, 2012

노안당의 가치를 높이는 빛 연출

노안당은 단청도 없이 단아하지만 다소 묵직한 고 건축물의 위엄을 지녔다. 이곳은 조선 후기의 사적 중에서 흥선대원군의 정치 활동 근거지로 유서 깊은 곳이다. 우리는 이곳에 화려한 빛이 아닌 담백한 빛을 격(格)이라는 관점으로 나타내고 싶었다. 그래서 빛의 연출기법 중 빛이 없는, 그러나 빛이 있는 연출기법을 사용했다. 운현궁은 서울의 중심에 있지만 주변 지역은 어둡게 조성되어 있었다. 서울 속에 작은 궁궐, 주변 어둠의 환경 속에 일부 시설이 밤에만 개방되는 특수성을 고려했다. 빛의 연출은 국가 문화유산을 안전하게 보호하는 최소한의 범위에서 빛의 디자인이 시작되었다.

Create Light that Enhances the Value of Noandang

Noandang stands simple but somewhat imposing ancient building without Dancheong. It is a historically significant place as it served as the political base for the activities of Heungseon Daewongun during the late Joseon Dynasty. We wanted to express the simple light, not the flashy light, as a kind of method called "gyeok" So, we used lighting techniques that have no light but still have light among the techniques of lighting production. Although Unhyeongung Palace is located in the heart of Seoul, the surrounding area was dimly lit. Considering the unique nature of having some facilities open only at night within a small palace in the heart of Seoul amid the dark environment, the lighting design began with the minimum scope to safely preserve the national cultural heritage.

이동식 무선 조명기구 개발로 국가 문화유산 보호

지금껏 사용하던 조명설치 방법은 조명기구를 설치하기 위해 땅을 파고 그 속에 전선과 조명기구를 설치하는 공사로 이루어졌다. 이 과정에서 국가 문화유산 고유의 모습이 일부 훼손되는 경우가 종종 발생했다. 처음부터 작은 전각을 비추기 위해 땅을 굴착해야 한다면 빛을 설치하지 않겠다고 생각했다.

여러 방법을 궁리하다가 이동식 전화기에서 아이디어를 얻었다. 이동이 가능하며 일정 시간만 사용할 수 있는 '이동식 무선 조명기구'를 개발했다. 일반적인 조명설치는 운현궁 꽃담에만 사용하고 노안당과 다른 전각(殿閣)에는 전기를 사용하지 않는 배터리(Battery)를 이용한 이동식 조명기구로 운현궁의 빛을 디자인해 냈다. 국가 문화유산에 대한 경관 훼손을 최소화하기 위해 개발한 충전용 조명기구는 국내 처음으로 국가 문화유산을 보호한 공로를 인정받아 특허청으로부터 경관조명용 특허를 받는 영광을 누렸다.

Protecting National Heritage by Developing Mobile wireless Luminaires

The traditional method of installing lighting consists of digging up the ground and installing wires and fixtures in it. In the process, some of the unique features of the national cultural heritage were often damaged. From the beginning, we realized that we would not install the light if we had to dig up the ground to illuminate the small pavilion.

After thinking about different options, I got an idea from a mobile phone. We developed a 'mobile wireless lighting fixture' that can be moved and used for a limited time. We designed the light of Unhyeongung Palace with battery-powered mobile lighting fixtures that do not use electricity for the Noandang and other pavilions, as conventional lighting installations are only used for the flower wall of Unhyeongung Palace. The rechargeable lighting fixture, which was developed to minimize landscape damage to national cultural heritage, was the first in Korea to receive a patent for landscape lighting from the Korean Intellectual Property Office in recognition of its contribution to protecting national cultural heritage.

배터리를 이용한 이동식 조명기구 빛의 테스트 | Light test using a portable lighting fixture powered by battery

낮은 조도의 빛 연출 테스트
Test of low-intensity lighting effects

운현궁의 꽃담길 호사

운현궁의 꽃담이 운현궁으로 들어오는 관람객을 반겨 맞이한다. 빨간색 꽃담에는 영세춘 수부강령 만세락(永世春 壽富康寧 萬歲樂)*이라고 쓰여 있다. 우리는 꽃담의 아름다운 문양과 빛깔을 밤 관람객에게 꼭 보여주고 싶었다. 빛을 어떤 방법으로 드러낼 것인지를 두고 다시 고민했다. 빛의 위치와 빛의 각도가 꽃담의 문양을 나타낼 수도 있고, 문양을 보이지 않게 할 수도 있는 것이다. 빛이 가까이에서 위를 비추면 빛의 그림자에 의해서 질감과 문양의 패턴이 보이고, 빛이 가까이 접근하면 그림자에 의해서 문양이 일그러지게 보일 수 있는 것이다.

현장의 다양한 빛 연출 테스트로 운현궁 꽃담 빛 창출

서울에 첫눈이 내린 날, 우리는 현장에서 이 부분을 직접 비추어 보면서 색감이 표현되는 실태를 파악하기 위해 밝은 조도의 빛과 낮은 조도의 빛을 번갈아 가며 비교했다. 조도가 높은 빛에서는 붉은 꽃담이 밝게 보이면서 붉은 색감이 모두 사라지는 현상을 볼 수 있었고, 빛의 조도를 낮췄을 때는 꽃담의 붉은 색감이 제대로 드러난다는 사실을 현장 빛 테스트를 통해서 알게 되었다.

첫째 문제는 황토벽의 느낌을 그대로 표현하는 방법이었다. 황토의 질감을 나타내기 위해서 조명기구와의 거리를 어느 정도 두어야 하는지 대해서는 조도 계산 프로그램에서 찾지 못했던 부분을 현장 테스트를 통해서 찾아냈다. 바로 꽃담의 모습을 비추는 조명기구와의 이격거리에 따라서 각기 다르게 보인다는 사실을 알게 된 것이다. 둘째는 관람자의 눈에 빛 기구가 보이지 않도록 하는 것이었다. 우리는 빛을 비추어 보았다. 꽃담을 바라볼 때 관람자 시각에 비추어 주는 빛이 먼저 보였다. 이렇게 넘쳐나는 빛을 차단하는 방법을 현장에서 고민하던 중에 새로운 아이디어를 찾게 됐다. 바로 조명기구 속에 또 하나의 조명기구를 넣어서 꽃담을 보는 관람자의 시선에서는 조명의 빛이 보이지 않도록 한 것이다.

다양한 현장의 테스트를 통해서 운현궁 꽃담의 정연한 빛을 창출해 냈다.

---

*영세춘 수부강령 만세락 (永世春 壽富康寧 萬歲樂): 영원토록 봄처럼 장수와 부귀를 누리고 강녕하시어 만년토록 즐거움이 함께 하소서.

Unhyeonggung Palace's Flower Wall

The flower wall of Unhyeonggung Palace welcomes visitors to Unhyeonggung Palace. The red flower wall is written with the words "Yeongsechun Subugangryeong Manserak"*. We wanted to show the beautiful patterns and colors of the flower wall to the night visitors. We thought again about how to reveal the light. The position of the light and the angle of the light could either reveal the patterns on the wall or make them invisible. When the light shines up close and from above, the texture and pattern of the patterns can be seen by the shadows of the light, and when the light moves closer, the patterns can be distorted by the shadows.

Creating light for Unhyeonggung Palace's flower wall by testing different light production on site

On the day of the first snowfall in Seoul, we alternated between high and low light to understand how the colors would appear in the field. In the high light, we could see that the red flower wall appeared bright and the red color disappeared altogether, but when the light was lowered, the red color of the flower wall was properly revealed.

The first problem was how to capture the feel of the clay wall. Through field tests, we found out what the illumination calculation program could not find: how far away the light fixtures should be from the clay wall to show the texture of the clay. Secondly, we realized that the appearance of the flower wall looks different depending on the distance from the luminaire that illuminates it. Secondly, we had to make sure that the light fixtures were not visible to the viewer. When looking at the flower wall, the light from the viewer's perspective came first. As we thought about how to block this overflowing light on site, we came up with a new idea. We put another luminaire inside the luminaire so that the light from the luminaire would not be visible from the viewer's point of view when looking at the flower wall.

Through various field tests, we were able to create an orderly light for the flower wall of Unhyeongung Palace.

---

*Yeongsechun Subugangryeong Manserak: May you enjoy long life, wealth and prosperity like spring forever, and may you be joyful for ten thousand years.

운현궁 이로당(二老堂) | Irodang, Unhyeongung Palace

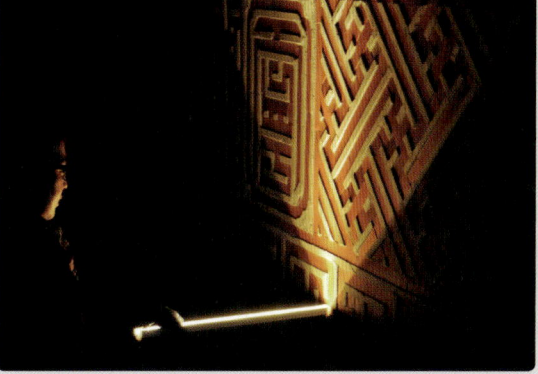

운현궁 꽃담 빛의 질감 테스트 | | Testing the light texture of the flower wall in Unhyeongung Palace

운현궁 이로당(二老堂) 빛의 디자인 | Light Design, Irodang, Unhyeonggung Palace

나에게 세상은 빈 그릇이다.
나는 빛으로 꿈과 환상을 담는다.

For me, the world is an empty vessel.
I capture dreams and fantasies with light.

한국은행 화폐금융박물관 빛의 디자인, 서울시 건축상 수상 | Light Design, Bank of Korea Museum of Money and Finance, Seoul Architecture Award, 2005

공간을 빛나게 하다:
빛과 소리로 충만한 풍경

MAKE THE SPACE SHINE:
LANDSCAPES FILLED WITH LIGHT AND SOUND

원주시 소금산 그랜드밸리 '2021 나오라쇼'
'2021 NAORA SHOW', WONJU SOGEUMSAN MOUNTAIN'S GRAND VALLEY IN WONJU CITY

거제시 장승포항, 동백꽃 붉은빛으로 피다
CAMELLIA BLOOMS IN RED, JANGSEUNGPO PORT, GEOJE CITY

창경궁 대춘당지 '물빛연화'
'MULBITYEONHWA', DAECHUNDANGJI POND, CHANGGYEONGGUNG PALACE

꿈을 담는 빛,
공간의 울림을 만드는 빛과 소리의 연출

The light of your dreams,
Light and sound to make a space resonate.

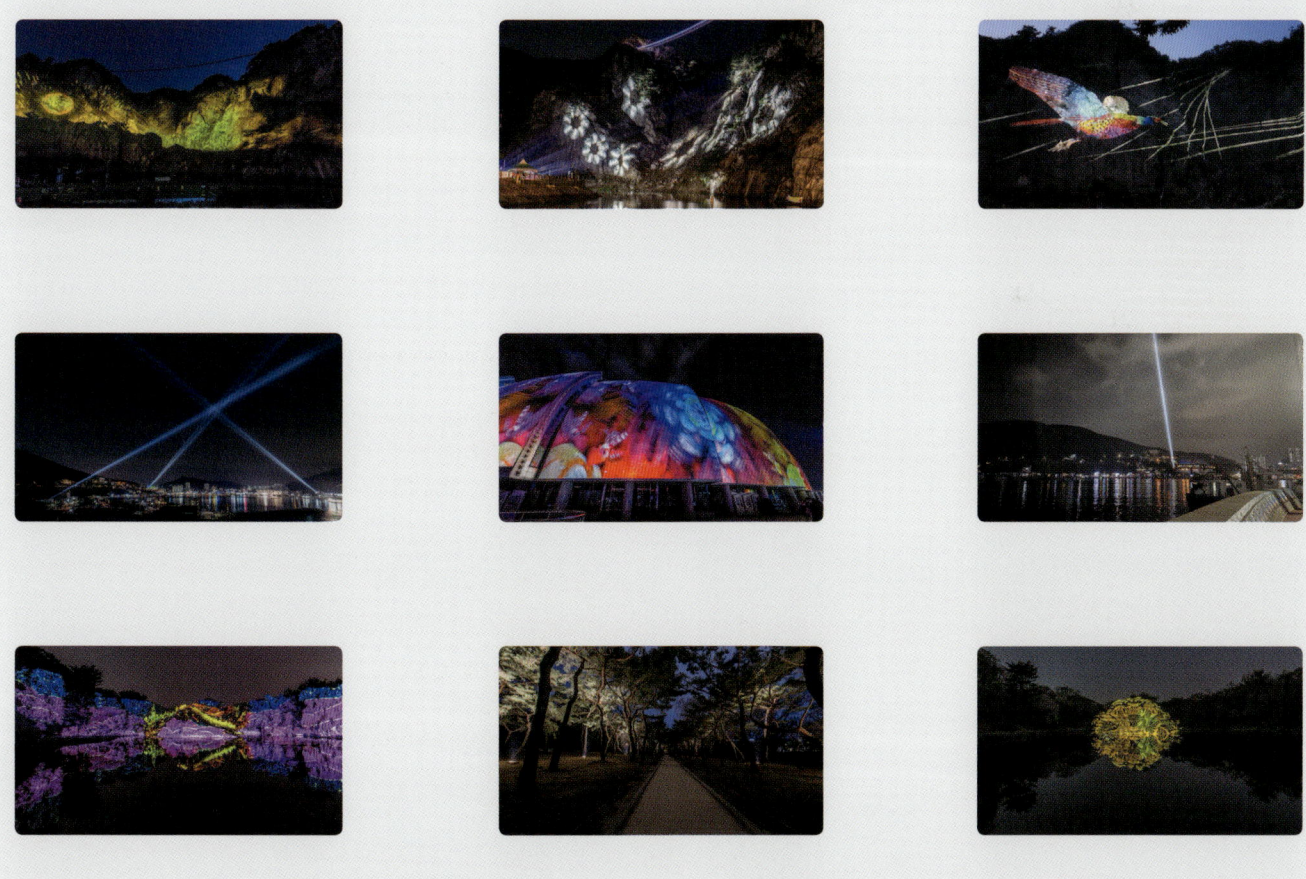

225

원주시 소금산 그랜드밸리 '2021 나오라쇼'

빛과 소리 영상과 분수, 하나의 감성을 담아내다. 총감독

2018년 1월 11일 강원도 원주시 간현관광지에 위치한 소금산 계곡에 새로운 출렁다리가 개장됐다. 그해 5월까지 이곳을 방문한 누적 관광객이 100만 명을 넘어섰다. 이렇게 간현국민관광단지가 관광명소로 발돋움한 데에는 원주시가 주도적으로 추진한 소금산 그랜드밸리 때문이다. 원주 간현관광지는 서울에서 광주원주고속도로를 이용해서 약 1시간 정도에 접근이 가능하고, 서울역에서 고속철도 KTX 강릉선 만종역을 이용하면 1시간 남짓으로, 접근성이 좋은 곳이다.

나에게는 도시와 공간은 하나의 무대(Stage)다

원주시에 사업 제안서를 낸 이유는 '밤의 가치 향상' 때문이었다. 원주 소금산 출렁다리가 100만 명 이상 방문하는 지역 명소가 되었지만, 계곡이라는 공간 특성상 주변보다 빨리 해가 지기 때문에 오후가 되면 출렁다리 입장을 서둘러 통제하고, 밤에는 안전상의 이유로 관광객 입장이 불가능했다.

우리가 현장에 도착했을 때는 이른 오후였는데, 벌써 사람들이 썰물처럼 다 빠져나가서 주변 가게와 음식점들이 문을 닫고 있었다. 순간, 나는 100만 명이 찾아오는 관광지에 방문객이 계속 머물게 할 수 있다면 지역 경제는 물론 원주시 경제까지 활성화될 수 있다고 확신했다.

우리가 제안한 '밤의 가치 향상'은 원주만의 차별화된 변화로, 원주 지역의 발전 동력이 될 수 있다는 판단이었다.

'2021 NAORA SHOW', WONJU SOGEUMSAN MOUNTAIN'S GRAND VALLEY IN WONJU CITY

Capturing one's Emotion with Light and Sound, Images and Fountains. General Director

On January 11, 2018, a new suspension bridge opened in the Sogeumsan Mountain Valley at Ganhyeon Tourist Area in Wonju, Gangwon-do. By May of that year, the cumulative number of tourists who visited the area exceeded 1 million. The development of the Ganhyeon National Tourist Complex into a tourist attraction is due to the Sogeumsan Mountain Grand Valley, which Wonju City took the lead in promoting. Wonju Ganhyeon Tourist Complex can be reached from Seoul in about an hour using the Gwangju-Wonju Expressway, and from Seoul Station, it takes just over an hour to Manjong Station on the KTX Gangneung Line.

For Me, the City and Space Exist as One Stage

The business proposal to the city of Wonju aims to enhance the nighttime appeal. The Wonju Sogeumsan Mountain Suspension Bridge has become a popular local attraction, drawing over 1 million visitors. However, due to the valley's location, the sun sets earlier than in the surrounding areas. As a result, the bridge is forced to close in the afternoon, and visitors are not allowed to enter at night due to safety concerns.

When we arrived at the site, it was early afternoon, and the crowds had already ebbed away, leaving the surrounding shops and restaurants closed. Immediately, I was convinced that if we could keep visitors in a tourist destination that attracts one million visitors, we could revitalize the local economy and even the economy of Wonju City.

Our proposal, 'Improving the value of the night', was a differentiated change that could become a development engine for the Wonju region.

길고 높은 산악 보도교 | Long and high mountain pedestrian bridge

원주시 소금산 그랜드밸리 출렁다리(길이 200m, 높이 100m, 폭 1.5m) | Sogeumsan Grand Valley Suspension Bridge, Wonju City (Length: 200m, Height: 100m, Width: 1.5m)

원주 간현의 소금산 이야기

소금산 그랜드밸리 빛 연출의 콘셉트는 '원주의 이야기'였다. 우리는 어디에도 없는 원주 소금산 그랜드밸리만의 스토리텔링으로 장소성을 드러낼 수 있다고 확신했다.

송강 정철이 「관동별곡(關東別曲)」에서 칭송한 간현 자연경관의 계곡과 암반에 새로운 빛과 소리를 통해서 발전하는 원주의 미래를 꿈꾸며 어둠의 간현에 새로운 빛을 꽃피우기로 했다.

The Story of the Sogeumsan Mountains of Ganhyeon County, Wonju

The concept for lighting the Sogeumsan Mountain Grand Valley was to tell the story of the Wonju City. We were convinced that we could reveal the sense of place through storytelling unique to the Wonju Sogeumsan Mountain Grand Valley.

We dreamed of a future Wonju that develops through new light and sound in the valleys and rock formations of the Ganhyeon County natural landscape praised by Song Kang Jeong-cheol in "Gwandongbyeolgok," and decided to bring new light to the dark Ganhyeon County.

원주시 소금산 그랜드밸리 '2021 나오라쇼' | '2021 Naora Show', Wonju City Sogeumsan Mountain Grand Valley

원주시 소금산 그랜드밸리와 '2021 나오라쇼' 가는 길

간현관광지 주차장에 들어서면 밝고 따뜻한 빛의 공간을 만나면서 우리의 발걸음은 한적한 간현암으로 향한다. 간현교의 다리 구간은 도로와 보행로가 혼용되어서 최대한 어린이들과 여성들이 편안하게 보행할 수 있도록 빛의 조도를 높이고 빛의 위치를 교량 아래쪽에 작게 설치했다. 빛이 보행 공간에만 머물도록 디자인한 것이다. 간현암으로 들어가는 삼산천교에서는 신비로운 느낌과 함께 변화되는 공간 이미지를 만나는데, 바로 간현암으로 가는 길의 시작점이다. 여기서 빛과 소리, 그리고 안개분수의 작은 물 입자에 맺히는 레이저의 불빛을 만난다. 삼산천교에서 간현암으로 들어가는 길을 빛의 동굴로 형상화했다. 신비로운 느낌을 주기 위해 교량 구조물에 레이저와 안개분수를 설치했다.

      삼산천변을 걸으며 숨어 있는 작은 소리를 듣는다. 수목에 비추어진 레이저가 만든 작은 반딧불이 빛의 모습과 수변 산책로 목재데크 길에 펼쳐진 빛의 꽃을 만날 수 있다. 그리고 빛과 영상의 홀로그램 연출로 신비한 세계를 접하며 간현암의 거북바위와 암석에 연출된 빛을 눈에 담으면서 간현암 계곡으로 들어간다.

      우리는 간현관광지 수중보를 지나 수중보 옆 보행교를 건너 간현암 앞에 이른다. 전반적으로 비교적 낮은 조도의 빛 속을 걸어서 이곳까지 왔다. 우리의 눈은 간현암 계곡의 어둠에 의해 더욱 어둡게 인지된다. 작은 태양광 조명기구의 빛으로 안전하게 관람객이 앉을 자리로 안내된다. 현실의 시간에서 이제 빛의 시간이다. 바로 원주, 소금산 그랜드밸리 '2021 나오라쇼'*다.

Directions to Sogeumsan Grand Valley and the '2021 Naora Show'

As we enter the parking lot of Ganhyeon Tourist Area, we are greeted by a bright and warm space of light, and our steps lead us to the secluded Ganhyeonam. The bridge section of the Ganhyeongyo Bridge is a mixture of road and sidewalk, so to make it as comfortable as possible for children and women to walk, the illumination level was increased and the light source was placed at the bottom of the bridge. The light is designed to stay in the walking space only. At Samsancheongyo Bridge, which enters Ganhyeonam, visitors encounter a mysterious and changing spatial image, which is the beginning of the path to Ganhyeonam. Here you encounter light, sound, and the glow of lasers on the tiny particles of water in the mist fountain. The path from Samsancheongyo Bridge to Ganhyeonam was visualized as a cave of light. To create a mysterious feeling, lasers and fog fountains were installed on the bridge structure.

      Walk along the Samsancheon Stream and listen to the small sounds that are hidden in the water. You can see tiny fireflies illuminated by lasers on the trees and light flowers on the wooden deck of the waterfront promenade. Then we enter the Ganhyeonam Valley, where the light and video holograms create a mysterious world and the turtle rocks and rock formations of Ganhyeonam are illuminated.

      We pass the underwater bridge of Ganhyeon Tourist Spot and cross the pedestrian bridge next to the underwater bridge to reach the front of Ganhyeonam. Overall, we walked here in relatively low light. Our eyes perceive the darkness of the Ganhyeonam Valley as even darker. The light of a small solar luminaire guides visitors safely to their seats. From the time of reality, it is now the time of light. This is Wonju City, Sogeumsan Mountain Grand Valley '2021 Naora Show'*

---

*간현관광지 야간코스「나오라쇼」: 「나오라쇼」는 나이트 오브 라이트(Night of Light)를 줄인 말이다. 2021년 10월에 시작된 아름다운 조명(삼산천교를 따라 레이저, 안개분수, 빛의 터널이 만들어내는 환상적인 빛)과 신비한 이야기를 영상화했다. 설화를 담은 미디어파사드는 간현암 계곡의 폭 250m, 높이 70m의 자연 암벽에 국내 최대 규모와 최고 화질의 빔프로젝트를 활용하여 원주 지역의 대표 설화 '은혜를 갚은 꿩' 등 다양한 영상 콘텐츠를 상영한다. 하천에 설치된 국내 최대 규모의 음악분수는 60m의 수직 물줄기와 함께 감미로운 음악 감상을 동시에 즐길 수 있는 간현관광지의 야간코스이다.

*"Naora Show," a nighttime course at a Ganhyeon sightseeing spot: "Naora Show" is short for Night of Light. Launched in October 2021, it features beautiful lights (lasers, fog fountains, and tunnels of light along Samsancheongyo Bridge) and mysterious stories. The storytelling media façade utilizes Korea's largest and highest-quality beam projector on a natural rock wall 250 meters wide and 70 meters high in the Ganhyeonam Valley to screen various video contents, including the representative story of Wonju City, "The Pheasant Who Repaid the Favor. The largest musical fountain in Korea installed on a river is a nighttime course at Ganhyeon Tourist Area where you can enjoy sweet music along with a 60-meter vertical stream of water.

삼산천교와 홀로그램 | Wonju Sogeumsan Mountain Samsancheon Bridge and Hologram

원주시 소금산 거북바위 | Wonju Sogeumsan Mountain Turtle Rock

자연에서 영감을 얻은 자연의 예술품 '나오라쇼'

간현암은 70m 높이에 두 개의 봉우리와 폭 250m 계곡의 거대한 모습을 38개의 빛(LED 960W/h 8ea, LED 450W/h 30ea)으로 위용을 드러낸다. 낮에는 볼 수 없고 느낄 수 없었던 간현암의 웅장함과 마주하게 된다.

빛이 일제히 소등되고, 어둠 속에서 울려 퍼지는 큰 소리는 나오라쇼의 시작을 알린다. 나오라쇼에서 빛보다 비중 있게 도입한 부분이 바로 소리였다. 간현암은 깊은 산 계곡에 있고, 이곳은 관광지였다. 밤에 크게 효과를 줄 수 있는 요소가 빛보다 소리라고 생각했다. 시작을 알리는 크고 긴 소리(9.1채널의 사운드 시스템)가 잠자고 있던 물의 빛 분수를 깨운다. 80m 길이의 긴 물빛이 음향의 선율에 맞추어서 다양한 빛으로 물기둥이 변화한다. 이제 간현암에 숨겨졌던 직선의 빛인 레이저 7개의 빛(30W/h 2ea, 22.4W/h 2ea, 16.8W/h 3ea)이 물과 만나면서 60m까지 분수의 수직 선형의 물길을 따라 빛이 흐른다. 크게 울려 퍼지는 소리, 그리고 생동감이 넘치는 분수의 모습과 선형의 빛인 레이저 연출로 간현의 밤이 화려한 빛으로 물든다. 관람객에게 강원도 원주의 대표적 설화 '은혜 갚은 꿩의 이야기'*가 간현계곡 자연 암반 스크린에 미디어파사드 영상으로 전개된다.

'Naora Show,' Artwork Inspired by Nature

Ganhyeonam is a 70-meter-high mountain with two peaks and a 250-meter-wide valley that is illuminated by 38 lights (8 LEDs 960W/h, 30 LEDs 450W/h). You will be confronted with the magnificence of Ganhyeonam that you cannot see or feel during the day.

The lights extinguish in unison, and a loud sound echoing in the darkness signals the beginning of the Naora Show. The sound is the most important aspect of the Naora Show, more so than the light. Ganhyeonam is located in a deep mountain valley, and it was a tourist destination. I thought that sound would have a greater effect at night than light. A loud and long sound (a 9.1-channel sound system) that announces the start of the show wakes up the dormant water light fountain. The 80-meter-long column of water changes into various colors of light in time to the melody of the sound. Seven lasers (30W/h 2ea, 22.4W/h 2ea, 16.8W/h 3ea), straight lines of light hidden in the Ganhyeonam, now meet the water, and the light flows along the fountain's vertical linear water path up to 60 meters. The loud sound, the lively appearance of the fountain, and the linear light of the lasers illuminate the night in Ganhyeon with colorful light. Visitors are treated to a media façade video of 'The Story of the Grateful Pheasant,'* a representative tale of Wonju, Gangwon-do Province, projected on a natural rock screen in the Ganhyeon Valley.

*길 가던 나그네가 뱀에게 먹히려던 꿩을 구해준다. 이번에는 나그네가 죽을 위기에 처하자, 그 꿩이 보은으로 종을 머리에 부딪쳐 종소리를 내는 바람에 죽을 위기에서 목숨을 건졌다는 보은(報恩)에 관한 전설이다. 전설에 따라 치악산(雉岳山)이 됐다.

*The Story of the Grateful Pheasant: A traveler saves a pheasant from being eaten by a snake. This is a legend about a pheasant that was saved from death by a traveler who struck it on the head with a bell to make it ring. As the legend goes, the mountain became Chiaksan Mountain.

간현암(250m×70m)에 미디어파사드로 연출된 원주 치악산 상원사 설화, 은혜 갚은 꿩의 전설 | The legend of the pheasant who repaid the favor, presented as a media façade on Ganhyeonam (250m×70m), Wonju Chiaksan Sanwonsa TempleTemple

나오라쇼 1000명 객석 | Naora Show with thousand seats

나오라쇼의 배경 간현암 | Ganhyeonam, Naora Show backdrop

미디어파사드, 사운드, 레이저, 분수, 경관조명이 하나로 연출된 강원도 원주시 소금산 그랜드밸리 '2021 나오라쇼'
'2021 Naora Show', Media facades, sound, lasers, fountains, and landscape lighting in the Sogeumsan Mountain Grand Valley in Wonju, Gangwon-do, Korea

간현 나오라쇼 스토리　　Kanhyeon Naorasho Story

1. 인트로 | Intro

시작을 알리는 음악소리 | Music to start the story

2. 선비의 여정 | The Journey of a Zen master

과거 시험을 보러 한양으로 향하는 선비의 새로운 도전과 희망을 나타내는 일출 | Sunrise signaling new challenges and hopes for the master as he travels to Hanyang to take the past exams

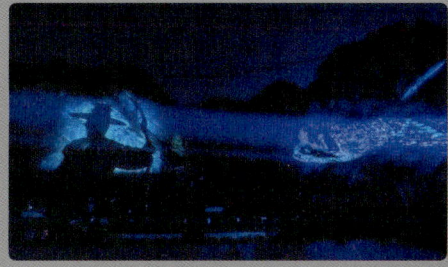

3. 위기의 꿩 발견 | Discovery of the Pheasant in Crisis

숲속에서 구렁이에게 잡아먹힐 위기에 처한 꿩을 구하는 선비 | In the forest, the monk saves a pheasant from being eaten by a pit bull.

4. 산중의 하룻밤 | Night in the mountains

날이 저물어 산속에서 하룻밤을 보내는 선비 | A monk spends the night in the mountains as the sun sets.

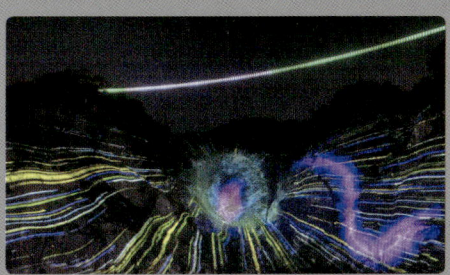

5. 구렁이의 복수 | The Revenge of the Beast

밤중에 나타난 구렁이가 선비를 위협 | A nighttime creature threatens the monk

6. 종소리의 구원 | Salvation by the Bells

상원사에서 들려오는 종소리로 구렁이가 선비를 풀어줌 | The sound of the bell from the senate chamber frees the monk from his captors.

7. 은혜 갚은 꿩의 희생 | The Sacrifice of the Grateful Pheasant

상원사에서 꿩이 종을 울려 선비를 구하고 죽음 | A pheasant rings the bell at the Senate House to save the monk and dies

8. 치악산의 유래 | Origin of Mount Chiaksan

꿩의 은혜를 기리기 위해 적악산이 치악산으로 이름이 바뀜 | In honor of Pheasant's grace, Red Mountain is renamed to Chiaksan Mountain

9. 선비의 과거 급제와 귀환 | Seonbi's past exams and return

한양에서 과거 시험에 합격한 선비가 꿩의 무덤을 찾아 감사의 마음을 전함 | A Zen master who passed his past exams in Hanyang visits Pheasant's grave to express his gratitude

세계적인 걸작 원주시 간현 '나오라쇼' 탄생

우리는 '나오라쇼' 빛 연출 디자인 설계를 진행하면서 간현관광지에 가장 크고, 가장 웅장하면서도 강원도 원주만의 이야기, 한국의 대표적 설화인 '은혜 갚은 꿩의 이야기'를 중심에 뒀다.

간현암 계곡의 미디어파사드는 Three Technic*에 담았다. 첫째 The Biggest. 국내 최대 규모의 영화 아이맥스(IMAX)** 스크린보다 7배가 더 큰 자연 암반을 스크린으로 사용해서 영상을 만들었다. 둘째 The Best. 기존 영상 해상도를 넘어서는 초고해상도의 이미지를 담고 있는 미디어파사드 영상 해상도로 제작했다. 셋째 The First. 야외공간에 빛과 9.1채널 사운드시스템, 레이저와 분수가 함께하는 암벽과 계곡에 연출된 최초의 미디어파사드이다. 이것이 우리가 자부심으로 연출한 '원주시 간현 나오라쇼'이다.

우리가 디자인한 나오라쇼의 빛 연출 설계를 마치고 나서 나는 원주시장으로부터 '원주시 간현 나오라쇼' 총감독 위촉장을 받았다. 2018년 12월부터 4년여의 기간에 걸쳐 영상을 제작하여 2021년 10월 1일에 간현 나오라쇼가 그랜드 오픈됐다. 나는 빛과 소리, 영상과 분수, 레이저가 만들어 내는 오케스트라의 지휘자였고, 나의 곡이 많은 사람을 위해 원주 소금산에 울려 퍼졌다.

웅장한 첫 공연이 끝나고 박수갈채가 터져 나왔다. 그동안 우리가 가꿔온 꿈을 현실로 만들어 낸 것이다. 빛의 디자인 담당자, 콘텐츠 회사, 각 시설 분야의 제작 설치를 맡은 회사, 원주시 관계자들, 그리고 마지막까지 우리의 능력을 믿고 격려를 아끼지 않았던 원주시장의 의지가 '원주시 간현 나오라쇼'를 탄생시킨 것이다.

A Global Masterpiece, Wonju Ganhyeon's Naora Show, is Born

In designing the light production for the Naora Show, we focused on the story of the largest and most magnificent attraction in the Ganhyeon tourist area, the story of Wonju, Gangwon-do Province, and a representative Korean folk tale, the story of the pheasant who repaid the favor.

The media facade of Ganhyeonam Valley was captured by Three Technic*. First, The Biggest, which uses a natural rock formation as a screen that is seven times larger than the largest movie IMAX** screen in Korea. Second, The Best, which uses ultra-high resolution images that exceed the existing image resolution of the media façade. Third, The First, this is the first media facade to be projected on a rock wall and valley with light, a 9.1-channel sound system, lasers, and fountains in an outdoor space. This is the 'Wonju Ganhyeon Naora Show' that we are proud to have produced.

After finishing the light production design of the Naora Show we designed, I received a commission from the mayor of Wonju to be the general director of the 'Wonju Ganhyeon Naora Show'. From December 2018, we produced the video over a period of four years, and on October 1, 2021, the Ganhyeon Naora Show was grandly opened. I was the conductor of an orchestra of light, sound, video, fountains, and lasers, and my music echoed through the Sogeumsan mountains of Wonju for many people.

After the magnificent first performance, applause erupted: the dream we had been cultivating had become a reality. The light designers, the content company, the company that installed the production in each facility, the Wonju officials, and the will of the mayor of Wonju, who believed in our abilities and encouraged us until the end, gave birth to the 'Wonju Ganhyun Naora Show'.

*Three Technic: The Biggest, The Best, The First
**아이맥스(IMAX): Image Maximum, 인간의 눈으로 볼 수 있는 각도의 한계까지 모두 채운 영상

*Three Technic: The Biggest, The Best, The First
**IMAX: Image Maximum, images filled to the limit of the angles that the human eye can see.

간현암에 미디어파사드로 연출된 은혜 갚은 꿩의 전설 | The Legend of the Grateful Pheasant as a Media Facade in Ganhyeonam

원주시 소금산 그랜드밸리 '2021 나오라쇼' | '2021 Naora Show', Wonju Sogeumsan Mountain Grand Valley

거제시 장승포항, 동백꽃 붉은빛으로 피다

CAMELLIA BLOOMS IN RED, JANGSEUNGPO PORT, GEOJE CITY

거제시 장승포항, 동백(冬柏)의 빛 연출

경상남도 거제시의 시화(市花)는 동백꽃이다. 거제시 장승포항은 조선소와 수산업이 발전하면서 1980년 장승포시로 승격됐다. 그러나 2011년 1월 부산에서 거제까지 운항하던 여객선의 뱃길이 폐쇄되면서 항구로서의 영화가 막을 내린다. 이에 따라 장승포항의 숙박시설과 음식점 등 다양한 생활문화 시설의 이용도가 급속하게 낮아졌다. 항구 주변에 거제문화예술회관이 위치하는데, 이곳에서는 매년 1월 1일 새해를 밝히는 새해맞이 행사가 열리는 거제시의 명소이며 주변에는 한국전쟁 당시 부모와 가족을 잃은 어린이들을 돌보아 주던 애광원이 자리하고 있다.

Light Production of Camellia, Jangseungpo Port, Geoje City, Korea

The city flower of Geoje City, Gyeongsangnam-do, is the camellia. Jangseungpo Port in Geoje City was promoted to Jangseungpo Port in 1980 as the shipyard and fishing industry developed. However, in January 2011, the ferry service from Busan to Geoje City was closed, ending its movie as a port. As a result, the use of various living and cultural facilities such as accommodation and restaurants in Jangseungpo Port has declined rapidly. The port is home to the Geoje Culture and Arts Center, where the city's New Year's Day ceremony is held every January 1, and nearby is Aegwangwon, where children who lost their parents and family members during the Korean War were cared for.

동백꽃 잎을 닮은 5각형의 거제시 장승포항 | Jangseungpo Port in Geoje City with a pentagonal shape resembling a camellia leaf

거제시의 시화(市花) 동백꽃 | Camellias, the city flower of Geoje City

2016년 빛의 디자인과 설계가 3단계로 시작

빛은 생명이다. 우리는 우리가 만든 빛이 거제시의 장승포항을 밝고 따뜻하게 밝혀주는 희망이 되어주기를 소망하는 마음으로 빛디자인에 나섰다. 2016년 빛의 디자인과 설계가 시작되었고, 2년이 지난 2019년 1월 1일에 1단계와 2단계, 2021년에 3단계의 빛디자인과 설계가 완성됐다.

우리는 빛을 디자인하기에 앞서 장승포항의 빛들을 살펴보았다. 빛을 유형별로 보면 주택가에서 스며 나오는 빛과 항구 주변에 있는 시설물에서 나오는 빛, 그리고 밤 항구의 선형을 만들어 내는 도로조명과 공원조명, 각종 상업시설물의 빛, 숙박시설의 빛 등 다양한 빛이 산재해 있었다. 한마디로 빛에 대한 위계와 체계가 없이 혼란스러웠다. 도로조명은 창백했고, 일부 공간은 어두워 접근이 어렵거나 심지어 무서운 곳도 있었다. 야간경관을 고려해서 만들어졌을 기존의 장승포 거제문화예술회관의 경관조명도 건축물의 전체적인 윤곽과 특징의 일부 모습만을 앙상하게 드러내고 있었다.

지역주민과의 대화를 통한 빛 연출 디자인

장승포 빛 연출에서 시급한 것은 주민들의 마음을 얻는 일이었다. 그래서 주변 상가와 숙박시설, 지역주민들의 다양한 이야기를 들었다. 듣는 중에, 귀에 들어오는 이야기가 있었다. "장승포항이 과거에는 부산을 오가는 중요한 항구의 역할을 했고, 이제 그 기능이 약해졌지만 아직 장승포항에서 유람선을 타고 외도와 해금강을 관광하는 사람들이 많다."는 것이다. 그리고 "항구 주변에 고급 리조트, 주변 펜션단지 등 숙박시설이 산재해 있는데, 밤이 되면 볼거리가 없어서 저녁이면 다른 지역으로 훌쩍 떠나버린다."는 것이다. 주민과의 대화 중에 우리는 새로운 가능성을 찾았다. 빛이 주민들에게 편리함과 안전을 제공하고, 장승포항에 특화된 빛을 연출해서 새로운 장승포를 창출해 내겠다는 꿈이었다.

Light and Design in 2016 Begins in Three Phases

Light is life. We set out to design the light with the hope that the light we created would be a hope to brighten and warm Jangseungpo Port in Geoje City. The design and construction of the light began in 2016, and two years later, on January 1, 2019, the design and construction of phases 1 and 2, and phase 3 in 2021, were completed.

Before designing the lights, we looked at the lights of Jangseungpo Port. By type of light, there were various types of light, including light from residential areas, light from facilities around the port, road and park lights that create the linear shape of the port at night, light from various commercial facilities, and light from accommodation facilities. In short, it was chaotic, with no hierarchy or organization to the light. The street lighting was pale, and some spaces were dark, making them inaccessible or even scary. Even the landscape lighting of the existing Jangseungpo Geoje Cultural Center for the Arts, which should have been designed with the nighttime view in mind, only revealed a small portion of the overall outline and features of the building.

Designing for Light through Community Conversations

The most urgent task in lighting Jangseungpo Port was to win the hearts and minds of the residents, so we listened to various stories from the surrounding shops, accommodations, and residents. While listening, one story stood out. "Jangseungpo Port used to be an important port for traveling to and from Busan, and although its function is now weakened, there are still many people who take cruise ships from Jangseungpo Port to sightsee in Oedo and Haegeumgang River." "There are accommodations such as luxury resorts and pension complexes scattered around the harbor, but there is nothing to see at night, so they leave for other areas in the evening." During our conversations with the residents, we realized a new possibility. We dreamed of creating a new Jangseungpo Port by creating a light specialized for Jangseungpo Port, where light provides convenience and safety for residents.

거제시 장승포항 동백의 빛 테스트 | Light test of camellias at Jangseungpo Port, Geoje City

미디어아트로 연출되는 동백꽃의 거제문화예술회관 | Geoje Cultural Arts Center with camellia flowers on the media art, 2021

항구 주변을 밝히는 빛의 마스터플랜 수립

장승포항을 총 4개 구역으로 구분하고, 각 공간의 성격이 특징적으로 드러나도록 장기적인 관점에서 도로조명의 색감과 밝기를 개선하고, 중기적으로 새롭게 조성되는 항구시설 일부가 공원으로 개발되는 광장, 항구의 수직적인 빛을 만들어 내는 민간 숙박시설 건물의 벽면 일부를 활용하는 빛 활용 계획을 세웠다.

기본계획을 수립하고, 주민들을 다시 만나 사업 설명에 나섰다. 모든 사업이 한꺼번에 이루어지기는 어렵다. 이제부터 장승포는 여러분의 응원 속에서 시간을 두고 연차적으로 함께 만들어 가야 한다. 혹여 첫 사업이 마음에 들지 않아도 마스터플랜에 의해 만들어 가는 사업을 믿고 적극적으로 지지하고 다음 단계로 나간다면, 마지막에는 큰 그림이 완성되어 새로운 가치의 장승포를 창출해 낼 수 있다는 마스터플랜을 설명했다.

새로운 빛 연출이 장승포 주민의 마음속에서 먼저 만들어지기 시작하면서, 새로운 빛들이 하나둘씩 장승포를 밝히기 시작했다. 단기적으로, 정박된 작은 선박을 비추어 주는 빛, 그리고 거제문화예술회관 굴곡진 벽면에 미디어파사드를 연출하는 방법으로 차츰 빛을 조성해 나갔다.

Creating a Master Plan for Light around the Harbor

The Jangseungpo Port was divided into four zones. A long-term plan was developed to improve the color and brightness of road lighting to characterize each space. In the medium term, a plaza will be developed as a park in part of the new port facilities, and a partial wall of a private accommodation building will create vertical light in the port.

After finalizing the general plan, the team held another meeting with the residents to explain the project. It is difficult to accomplish everything at once. Even if you're not thrilled about the initial project, if you believe in and actively support the project outlined in the master plan and take the next steps, the overall vision will eventually be realized, creating new value for Jangseungpo Port.

As the concept of new lighting took shape in the minds of the residents of Jangseungpo Port, new lights began to illuminate the area one by one. In the short term, light was gradually introduced by illuminating small ships anchored and by creating a media facade on the curved wall of the Geoje Cultural Arts Center.

미디어아트로 연출되는 동백꽃이 피는 거제문화예술회관 | Camellia flowers blooming on the media art of Geoje Arts Center

빛과 영상으로 장승포항을 동백꽃으로 물들이다

하늘에서 장승포항을 내려다보면 5각형의 동백꽃 꽃잎처럼 보인다. 빛의 콘셉트는 거제시의 시화(市花) 빨간 동백꽃 형상화다. 빛이 동백 꽃잎이 되고, 항구는 온통 붉은 동백꽃의 빛으로 물든다. 거제문화예술회관은 빛과 영상으로 피어나는 동백꽃을 감상하는 곳이다.

우리는 많은 사람이 거제시 장승포항에 와서 행복을 이야기하고, 다시 사랑을 담아가는 곳으로 새롭게 피어나기를 기원하는 마음으로 빛 연출을 계획했다. 거제시 애광원을 설립한 사람들의 따뜻한 마음처럼 장승포항에 다시 활기가 샘솟기를 기원하면서.

Camellia Flowers Color Jangseungpo Port with Light and Video

When you look down on Jangseungpo Port from the sky, it looks like the petals of a pentagonal camellia flower. The concept of the light is the red camellia flower, the city flower of Geoje City. The light becomes camellia petals, and the harbor is bathed in red camellia light. The Geoje Cultural Arts Center is a place where you can watch the camellia bloom with light and video.

We planned the light production with the hope that many people will come to Jangseungpo Port in Geoje City to talk about their happiness and that it will bloom anew as a place where they go to make love again. We hope that Jangseungpo Port will be revitalized again, just like the warm hearts of the people who founded Aegwangwon in Geoje City.

4km에서 조망되는 장승포의 빛 | The light of Jangseungpo, seen from 4 kilometers away

매 정시 5분간 피어나는 장승포항 동백의 빛 연출 | Light display of camellias blooming for 5 minutes every hour, at Jangseungpo Port

창경궁 대춘당지 '물빛연화'

서울의 다섯 개 궁궐 중 유일하게 자연형 호수를 간직한 창경궁

창경궁은 사적 제123호로, 1483년 세조비 정희왕후, 예종비 안순왕후, 덕종비 소혜왕후 세 대비를 모시기 위해 옛 수강궁터에 창건된 궁궐로서, 1592년 임진왜란으로 모든 전각이 소실되었으나 일부 전각만이 재건됐다. 이후 근대시대를 거치면서 창경원으로 위상이 격하되었고, 궁궐의 많은 공간이 변형되면서 큰 호수인 대춘당지와 작은 호수 소춘당지 일부가 남게 되었다.

국가 문화유산을 관리하는 국가유산청에서는 국가 문화유산 보호를 넘어 활용 방안에 대해 숙고한 끝에 국가 문화유산을 공개하는 방안을 표방했다. 사람들에게 국가 문화유산의 가치를 느끼게 해주자는 것이다. 한정된 인원에게 밤시간, 일정 공간에 한해 국가 문화유산이 담고 있는 고유의 풍광을 외부에 개방키로 한 것이다. 이에 따라 궁중문화축전, 달빛기행, 별빛야행 등 밤시간에 국가 문화유산을 개방하는 다양한 행사로 발전했다. 지금까지 진행된 행사는 행사 기간에 이벤트성으로 일부 시설에 국한하여 가설(假設)되다 보니 낮에는 여러 시설물이 과도하게 노출되고, 관람객의 이동에도 안전 문제가 발생할 소지가 있었다.

우리는 창덕궁 달빛기행에 시각의 순응이라는 개념으로 빛을 새롭게 디자인해 내었다. 오랜 시간 동안 창덕궁 주요 지점에서 빛 테스트를 통해 전각의 형태와 주변 조경식재가 지닌 풍광에 따라 창덕궁의 분위기가 각기 다르게 느껴진다는 사실을 알게 되었다. 이를 고려하여 최소한의 빛 연출로 창덕궁의 동쪽 궁궐인 창경궁 밤의 풍광을 드러내기 위한 빛 연출 계획에 들어갔다.

'MULBITYEONHWA', DAECHUNDANGJI POND, CHANGGYEONGGUNG PALACE

Changgyeonggung Palace, the only One of Seoul's Five Palaces to Have a Natural Lake

Changgyeonggung Palace, Historic Site No. 123, was built in 1483 on the site of the former Suganggung Palace to honor the three great dowagers, Queen Jeonghee, Queen Ansun, and Queen Sohye, but all the buildings were destroyed during the Imjin War in 1592, and only some were rebuilt. The palace was later downgraded to a Changgyeongwon during the modern era, and many of the palace's spaces were transformed, leaving behind the large pond, Daechundangji Pond, and parts of the smaller pond, Sochundangji Pond.

The National Heritage Administration, which is responsible for managing national cultural heritage, has been considering ways to make better use of national cultural heritage beyond just protecting it. They have decided to make it accessible to the public in order for people to appreciate its value. As a result, they have chosen to open certain national cultural heritage sites to a limited number of people at night. This has led to the organization of various nighttime events at these sites, including court culture festivals, moonlight walks, and stargazing nights. Previously, events were temporary and limited to a few facilities during the event period, which meant that various facilities were overly exposed during the day and posed safety concerns for visitors.

We redesigned the light for the moonlight tour of Changdeokgung Palace with the concept of visual adaptation. Through years of light testing at key points in Changdeokgung Palace, we realized that the palace feels different depending on the shape of the main hall and the surrounding landscape plantings. Taking this into consideration, the lighting plan was designed to reveal the nighttime scenery of Changgyeonggung Palace, the eastern palace of Changdeokgung Palace, with minimal light production.

빛의 디자인, 창경궁(昌慶宮) 명정전(明政殿) 국보 제226호 | Light Design, Myeongjeongjeon Hall, National Treasure No. 226, Changgyeonggung Palace, 2024

창경궁의 물빛으로 이야기를 담는다

조선왕조 오백 년의 고결한 국가 문화유산 창경궁은 조선 왕실 여성의 공간이며, 조선시대 새로운 혁신을 추구했던 정조대왕이 머물던 곳이다. 정조대왕이 이곳에서 사랑하는 어머니 혜경궁 홍씨의 회갑연을 기념하여 창경궁의 정문인 홍화문 밖에 친히 나가서 가난한 백성들에게 쌀을 나누어 주는 애민사상(愛民思想)을 몸소 실천한 역사적인 장소이다.

정조대왕은 경춘전(景春殿)에서 탄생하여 영춘헌(迎春軒)에서 승하했다. 창경궁 '물빛연화'는 이런 창경궁이 품고 있는 조선 왕실의 우아하고 온화한 정을 대춘당지(大春塘池)의 물결과 주변의 자연풍광에 담아 창경궁 빛의 8경을 만들었다.

Capturing Stories with the Colors of Changgyeonggung Palace's Waters

Changgyeonggung Palace, a noble national cultural heritage from the five hundred years of the Joseon Dynasty, is a space for the women of the Joseon royal family and the residence of The Great King Jeongjo, who pursued new innovations during the Joseon Dynasty. This is the historic place where King Jeongjo personally practiced his philosophy of loving people by distributing rice to the poor outside Honghwamun Gate, the main gate of Changgyeonggung Palace, in honor of his beloved mother, Queen Hyeokgyeonggung Hong.

The Great King Jeongjo was born in Gyeongchunjeon and died in Yeongchunheon. Changgyeonggung Palace's 'Mulbityeonhwa' captures the graceful and gentle spirit of the Joseon royal family in the waves of Daechundangji Pond and the natural light of the surrounding area to create the Eight Scenes of Changgyeonggung Palace.

창덕궁과 창경궁을 묘사한 궁궐의 동궐도(東闕圖) | Painting of Donggwoldo, depicting Changdeokgung and Changgyeonggung Palace

창경궁 빛의 8경을 만나다

'창경궁(昌慶宮) 물빛연화'는 2021년에 시작된 사업으로, 「2021년 창경궁 빛 연출 기본계획」을 수립하고 3단계 사업을 연차별로 디자인과 설계, 총감독의 감리를 진행하여 2024년 창경궁 '물빛연화 빛의 8경'이 4년 만에 완성됐다.

      '물빛연화'는 창경궁으로 들어설 때부터 마지막 나오는 순간까지 전체적인 빛의 위계성에 대해서 구상하고 기존의 빛과는 다른 이미지를 보여주도록 계획했다. 바로 관람객에게 빛과 소리의 감성의 울림을 줄 수 있도록 빛과 영상을 하나의 톤으로 만들어냈다.

창경궁의 탐방로, 빛이 되어 현재와 과거, 미래를 담아내다

창경궁은 대한민국 수도 서울 한가운데 위치하는 궁궐로, 주변 도시지역에서 유입되는 조명과 소음이 발생하는 곳이다. 우리는 창경궁 고유의 어둠을 느낄 수 있는 공간을 찾으려고 노력했다. 창경궁만의 전각과 물길, 정원과 창경궁 대온실의 모습을 정연하게 보고 느낄 수 있는 새로운 동선을 찾아냈다. 평상시에는 닫혀있던 창경궁 중심에 있는 탐방로를 창경궁 '물빛연화' 관람의 동선으로 설정했다.

Meet the 8 Scenes of the Changgyeonggung Palace Lights

The 'Changgyeonggung Palace Mulbityeonhwa' project began in 2021, and after establishing the 'Basic Plan for Changgyeonggung Palace Light Production in 2021,' the three phases of the project were designed, constructed, and supervised by the general director, and the 'Eight Scenes of Changgyeonggung Palace Water Light Illumination' in 2024 was completed in four years.

      From the moment you enter Changgyeonggung Palace to the moment you leave, the entire hierarchy of light was conceived and planned to show a different image from the existing light. Light and image were created in a single tone to give the viewer the emotional impact of light and sound.

Changgyeonggung Palace's Pathways Become Lights, Capturing the Present, Past, and Future

Changgyeonggung Palace is a palace located in the center of Seoul, the capital of South Korea, with light and noise coming from the surrounding urban area. We tried to find a space where we could feel the darkness inherent in Changgyeonggung Palace. We found a new pathway that allowed us to see and feel the palace's unique pavilions, waterways, gardens, and the Changgyeonggung Palace Greenhouse in an orderly fashion. The pathway in the center of Changgyeonggung Palace, which is normally closed, was set up as the route for the Changgyeonggung Palace 'Mulbityeonhwa' exhibition.

진입로 소나무 숲 디밍 테스트 | Dimming test of the pine tree forest at the entrance

대춘당지(大春塘池) 수면 반사 테스트 | Reflection test on the surface of Daechundangji Pond

소춘당지(小春塘池) 빛 연출 에이밍 | Aiming light production, Sochundangji Pond

명정전(明政殿) 빛 연출 에이밍 | Aiming light production, Myeongjeongjeon

창경궁 대온실 빛 연출 테스트 | Greenhouse light performance test, Changgyeonggung Palace

빛이 흐르는 창경궁(昌慶宮)의 소나무 숲길

제1경 창경궁 진입로, 대화할 화(話)를 만난다

창경궁 숭지문을 나서서 소나무 숲길로 들어선다. 소나무가 빛과 함께 호흡하며 궁궐로 들어선다. 수많은 나라의 의지들이 반딧불이처럼 소나무에 모이고, 모여든 자리마다 횃불이 밝혀지고 새역사를 향한 빛의 주단이 역동적으로 대춘당지로 모여든다. 공간을 울리는 북소리와 빛들의 움직임이 옛 창경궁의 영화 속으로 우리를 불러들인다.

    빛의 진입부인 소나무 숲길에 낮게 흐르는 부드러운 음악을 따라 앞으로 흘러 들어가는 변화되는 빛을 연출했다. 소나무 그루마다 빛을 하나씩 비추어 가면서 빛의 느낌을 관찰하여 조명 설치 위치와 방법을 찾아서 큰 빛의 흐름으로 표현했다. 탐방객이 걸어가는 속도에 맞추어서 빛과 소리가 생동감 있게 변화하도록 빛을 연출했다. 우리는 이런 빛과 소리의 이동으로 신비한 밤의 분위기를 창출해 낼 수 있다는 사실을 알게 됐다.

제2경, 창경궁(昌慶宮)의 백송(白松), 채색할 화(畵)를 느낀다

대춘당지를 지나 수변 산책로를 걷는다. 창경궁에서 오랜 시간을 함께한 백송(白松) 3형제와 마주한다. 빛의 물결이 흩날리고 뿌려져서 다시 태어나는 백송의 모습에서 창경궁의 단아한 모습이 빛을 입고 섰다. 백송은 창경궁이 간직한 품격을 나타내며, 오랜 세월 변치 않고 빛나는 창경궁의 상징이다.

The Pine Forest Paths of Changgyeonggung Palace, Illuminated by Light

Scene 1, Changgyeonggung Palace Entrance, Meet Hwa for Conversation

Leave the Sungjimun Gate of Changgyeonggung Palace and enter the pine forest. The pine trees breathe with light as you enter the palace. The wills of countless nations gather in the pine trees like fireflies, torches are lit at each gathering, and the main body of light for a new history dynamically gathers at Daechundangji Pond. The sound of the drums echoing through the space and the movement of the lights draws us into the movie of the old Changgyeonggung Palace.

    At the entry point of the light, the pine forest path, the soft music played low, creating a changing light that flows forward. By shining a light on each pine tree one by one and observing the feeling of the light, we found the location and method of installing the lights to create a large flow of light. We designed the light so that the light and sound would change in a lively way according to the speed of the visitors walking. We realized that this movement of light and sound could create a mysterious nighttime atmosphere.

Scene 2, The White Pine of Changgyeonggung Palace, Feeling the Painting to Paint

Walk along the waterfront promenade past Dachendangji Pond. You come face to face with the three Baeksong(white pine) brothers, who have spent many years at Changgyeonggung Palace. As the waves of light scatter and reborn, the elegant appearance of Changgyeonggung Palace is bathed in light. The white pine represents the dignity of Changgyeonggung Palace, and is a symbol of its timelessness.

대춘당지 진입로 소나무길과 백송(白松) | Entrance to Daechundangji Pond, featuring the pine tree path and white pines

제3경, 창경궁 대춘당지, 피어날 화(花)를 듣고 바라본다
대춘당지 앞에서 빛나는 호수의 물빛을 바라본다. 2024년 물빛연화에서는 '애민사상'의 시각화로, 왕의 백성에 대한 사랑의 마음이 큰 파도로 솟구치며 일어선다.

Scene 3, Daechundangji Pond, Changgyeonggung Palace, Listening to and Watching the Blooming Flowers
A view of the shining lake in front of Daechundangji Pond. In the 2024 Water Light Painting, a visualization of "Aeminsasang" the king's love for his people rises up in great waves.

<서장> 봄의 연못, 이야기의 시작
- 우아한 날갯짓으로 대춘당지 위를 평화롭게 날아다니는 나비와 사방에 가득한 복숭아나무
- 춘당 이야기의 시작을 알린다.

<Prelude> Spring Pond, the beginning of a story
- Butterflies fluttering gracefully over Daechundangji Pond and peach trees everywhere
- Mark the beginning of the story of Chundang.

<제1막> 봄기운에 설레는 여인의 마음
- 춘당에 포근한 봄바람이 불어오고,
- 숲을 산책하며 나무와 교감하는 여인의 기쁨.
- 여인들의 공간인 창경궁의 아름다운 풍광이 눈앞에 펼쳐진다.

<Act 1> A woman's heart filled with excitement for spring
- A warm spring breeze blows through the Chundang,
- Woman takes a walk in the forest to commune with the trees.
- The beautiful scenery of Changgyeonggung Palace, a space for women, unfolds before your eyes.

<제2막> 왕의 고뇌
- 나라의 평안과 백성의 평화로운 삶을 근심하는 왕
- 봄의 연못 위 정자에서 여전히 고뇌하는 왕의 모습

<Act 2> The King's Anguish
- The king worries about the peace of his country and the peaceful lives of his people.
- The king is still anguished in his pavilion over the spring pond.

<제3막> 홍화(弘化)의 꿈
- 만백성의 안녕을 기원하는 연희가 열리는 홍화문.
- 아름다운 무희들의 춤은 나라의 안녕과 임금님의 만수무강을 기원한다.

<Act 3> Honghwa's Dream
- Honghwamun Gate is the site of a festival to pray for the well-being of the people.
- The beautiful dancers dance to pray for the well-being of the nation and the king's long life.

<제4막> 태평국을 꿈꾸다
- 나라를 지키는 다섯 개의 봉우리와 소나무가 솟아오르며, 그 앞에서 활기차게 휘몰아치는 파도가 춘당지 수면을 수놓는다.

<Act 4> Dreaming of the Kingdom of Taepyeongguk
- The five peaks and pine trees that protect the country rise up, and in front of them, the lively waves of Chundangji Pond bathe the water.

<제5막> 태평성대, 용의 승천
- 태평성대, 왕의 기원은 용이 되어 승천하고,
- 세상은 황금빛 들녘으로 태평성대(太平聖代)를 맞이한다.

<Act 5> Taepyeongseongdae, Ascension of the Dragon
- Taepyeongseongdae, the origin of kings, ascends as a dragon,
- and the world welcomes Taepyeongseongdae with golden fields.

한국 최초의 서양식 온실, 창경궁(昌慶宮) 대온실 | The Great Greenhouse of Changgyeonggung Palace, the first Western-style greenhouse in Korea

제4경, 빛으로 물든 창경궁 대온실, 빛날 화(華)의 물빛으로 승화하다

창경궁 대온실 앞마당은 흙길이다. 넓게 펼쳐진 부드러운 흙의 공간에 물빛이 서서히 퍼져나간다. 빛은 어둠 속에서 물빛 너울을 만들고, 대온실을 바라본다. 대온실의 빛 연출은 외부에서 상부구조에 일부 시간만 빛을 투사하는 새로운 빛 연출기법으로 전환하여 빛을 디자인했다. 대온실 내부 식물이 빛에 의해 영향이 미치지 않도록 외부에서 온실구조 상층부에만 빛을 투사하는 방식으로, 대온실 유리구조의 특성을 실루엣으로 살려냈다.

Scene 4, The Large Greenhouse of Changgyeonggung Palace, Drenched in Light, Transformed into a Shining Hwa

The courtyard in front of Changgyeonggung Palace's large greenhouse is a dirt path, and the color of water gradually spreads across the wide expanse of soft earth. The light creates a watery hollow in the darkness and looks toward the greenhouse. The light design for the greenhouse was a transition to a new light production technique that projects light from the outside onto the superstructure only part of the time. In order to prevent the plants inside the greenhouse from being affected by the light, the light is only projected from the outside onto the upper layer of the

제5경, 비밀의 정원 소춘당지, 예쁠 화(姡)의 빛으로 물의 숨결을 담다

우리는 창경궁 후원 소춘당지로 향한다. 이곳은 창경궁에서 가장 정적인 공간으로, 숨어 있는 비밀 정원과 같은 곳이다. 높은 곳에서 나무 사이로 소춘당지의 낮은 수면을 조망하게 된다. 수면의 빛이 주변 자연에 투영되도록 빛을 연출했다. 정조대왕의 애민 사상을 빛의 영상으로 담아냈다.

제6경, 빛으로 수놓은 길, 화목할 화(和)의 빛나는 빛의 길

아름다운 별빛이 드리워진 선경, 이상향의 길이다. 역사의 시간을 거쳐 빛으로 수 놓은 길을 지나 미래를 향하는 빛의 통로에 들어선다. 환상적인 빛은 사라졌던 반딧불이가 따라붙어 이상향으로 들어가는 빛의 물결이 빠르게 출렁인다.

제7경, 홍화의 물빛, 화합할 화(和)의 의미를 담아 찬란한 궁을 빛으로 새기다

창경궁의 마지막 탐방로에서 금빛 궁의 흉배 문양과 금빛의 한글을 만난다. 역경을 딛고 일어서기를 염원하는 빛의 물결에 화합과 소생의 기운이 솟는다. 빛이 새로운 물길을 만들고 언덕을 오를 때 멀리 보이는 나무와 나무의 그림자가 우리를 몽환적인 세계로 이끈다.

제8경, 영원한 궁, 빛이 될 화(化), 창경궁의 아름다운 메시지를 전해준다

붉게 물든 홍화의 물빛 속에서 창경궁을 기억하는 감성의 문장들이 빛으로 살아나면서 현실 세계로 돌아온다.

Scene 5, Secret Garden of Sochundangji Pond, Capturing the Breath of Water with the Light of a Beautiful Hwa

We head to the Sochundangji Pond at the rear garden of Changgyeonggung Palace. This is the most tranquil area of the Sochundangji Pond, like a hidden secret garden. From high above, we can see through the trees to the water below. The light is arranged so that the light from the water is reflected on the surrounding nature. The light is designed to reflect the surrounding nature.

Scene 6, The Path Embroidered with Light, The Shining Path of Light of Reconciliation and Harmony

A beautiful starlit vista, leading to an ideal world. As you travel through time, you pass along a path adorned with light and enter a bright passage to the future. The dazzling light is accompanied by fireflies that had vanished, and the waves of light entering the ideal world are swiftly moving.

Scene 7, Inspiring Water Color, Carving a Radiant Palace with Light, Meaning Harmony and Unity

On the final tour of Changgyeonggung Palace, you'll encounter the golden palace's graven image and golden Korean script. A sense of unity and revitalization rises in the streams of light that wish to rise above adversity. As the light creates new paths and climbs the hill, the distant trees and their shadows draw us into a dreamlike world.

Scene 8, The Eternal Palace, Hwa, the Light of the World, Delivers the Beautiful Message of Changgyeonggung Palace

In the reddish color of the praying for the well being of the people, sentimental sentences remembering Changgyeonggung Palace come alive with light and return to the real world.

제5경 비밀의 정원 소춘당지 | 5th View: Secret Garden, Sochundanji Pond

제6경 빛으로 수놓은 길 | 6th View: Path embellished with light

제7경 홍화의 물빛 | 7th View: Water light of Honghwa

제8경 영원한 궁 | 8th View: Eternal palace

빛 연출로 소중한 국가 문화유산을 널리 알리다

창경궁 빛 연출은 국가 문화유산에 대한 새로운 인식 변화로, 국가 문화유산의 활용 측면에서 시사하는 바가 자못 크다. 2024년 창경궁 물빛연화가 새로운 변화를 알리는 물결의 시작점이자 변화를 선도하는 새로운 빛 연출의 물길이 되기를 기대해 본다.

지역 고유의 환경을 고려한 빛 연출

나는 아름다운 꽃을 볼 때마다 깊은 땅속 어둠에 무슨 마력(魔力)이 있어서 이토록 화사한 꽃을 피워낼까, 경이롭기만 했다. 그런데 꽃이 환경에 따라 서로 다른 속살로 피어난다는 사실을 새롭게 알았다. 일찍이 신토불이(身土不二)라는 말이 있거니와, "몸과 자신이 태어난 땅은 둘이 아니고 하나"라는 뜻으로, 자신이 태어난 땅에서 나온 먹거리가 제 몸에 잘 맞는다는 뜻이다. 결국, 그 지역의 풍토에 따라 꽃이 다르게 피어난다는 뜻이기도 하다. 지역 환경에 따라 고유문화가 꽃피는 것이다. 나는 이제 지역성(地域性, locality)을 고려한 빛 연출에 대해 말해야 할 것 같다. 오늘의 우리 문화가 중앙과 세계의 문화를 중심에 뒀다면, 이제 지역의 고유 문화에 눈을 돌려야 한다.

지역의 특성을 밝히는 빛의 디자인

우리 지역이 가지고 있는 지금의 모습으로는 현재 그 이상을 넘어설 수가 없다. 그러려면 새롭게 보아야 한다. 지역을 밝히는 새로운 감성의 빛 연출은 그 지역 사람에게 행복을 느끼도록 하고, 사람들의 방문과 소통으로 지역경제가 발전하고, 새로운 고용 창출로 지역이 부흥하는 것이다. 홍콩과 리옹이 과감하게 발상을 전환했고, 원주시 소금산 그랜드밸리가 획기적으로 '2021 나오라쇼'를 도입했다.

지역 공간이 빛 연출로 활성화되면 지역의 명소(名所)를 넘어 세계의 명소로 널리 알려질 것이다. 지역은 세계의 변화를 받아들이고, 함께 변화하면서 앞서가야 한다. 변화에 대한 꿈이 곧 지역의 미래가 되기 때문이다.

Promote Our Precious National Cultural Heritage through Light Production

The illumination of Changgyeonggung Palace is a new change in perception of national cultural heritage, which has great implications for the utilization of national cultural heri-tage. It is hoped that the 2024 Changgyeonggung Palace's Mulbit-yeonhwa will be the beginning of a new wave of change and the beginning of a new wave of light production leading to change.

Light for Your Local Environment

Whenever I see a beautiful flower, I wondered what kind of magic in the deep darkness of the earth could produce such a colorful bloom. But then I realized that flowers bloom in different ways depending on their environment. There is an old saying, 'Shintobului' which means 'your body and the land you were born in are not two, but one,' and it means that the food that comes from the land you were born in fits your body well. In other words, different flowers bloom differently depending on the local climate. I think it is time to talk about light production that considers locality. We must now turn our attention to local unique cultures so that our culture today can become the center of the world.

Designing Light to Illuminate the Characteristics of the Region

We cannot exceed our current state with the current situation in our region. In order to achieve this, we need to change our perspective. Introducing a new approach to lighting that brightens up the area will bring joy to the local community, stimulate the local economy by attracting visitors and fostering communication, and rejuvenate the region by generating new employment opportunities. Hong Kong and Lyon have taken a bold step forward, and Wonju Sogeumsan Mountain Grand Valley in Wonju City has introduced the '2021 Naora Show' in a groundbreaking way.

When local spaces are activated by light production, they will be widely recognized as not only local attractions, but also global attractions. Regions must embrace the changes in the world and change with them to stay ahead of the curve. The dream of change is the future of the region.

창경궁 물빛연화의 무대가 된 대춘당지의 숲 | The Forest of Daechundangji Pond, stage of Mulbityeonhwa, Changgyeonggung Palace, 2024

빛을 밝히다:
장소성을 담아내는 영롱한 빛

LET THE LIGHT SHINE:
BRILLIANT LIGHT THAT CAPTURE A SENSE OF PLACE

석촌호수, 물의 모습을 빛과 소리로 담다
SEOKCHON LAKE, CAPTURING THE APPEARANCE OF WATER THROUGH LIGHT AND SOUND

서해에서 불어오는 바람을 빛으로 느끼다
FEEL THE BREEZE BLOWING FROM THE WEST SEA THROUGH LIGHT

제주도 서귀포 생태공원 달빛과 별빛을 품다
EMBRACING MOONLIGHT AND STARLIGHT, SEOGWIPO ECOLOGICAL PARK, JEJUDO ISLAND

서울, 빛의 혈(血)을 발견하다
SEOUL, DISCOVERING THE VEINS OF LIGHT

나의 감성은 자연의 품 속에서 고요히 깨어난다.
감성이 스민 빛은 공간 속에 스며들어, 새로운 가치를 피워 낸다.

My emotions quietly awaken in the embrace of nature.
The light, imbued with emotion, seeps into the space and brings forth new value.

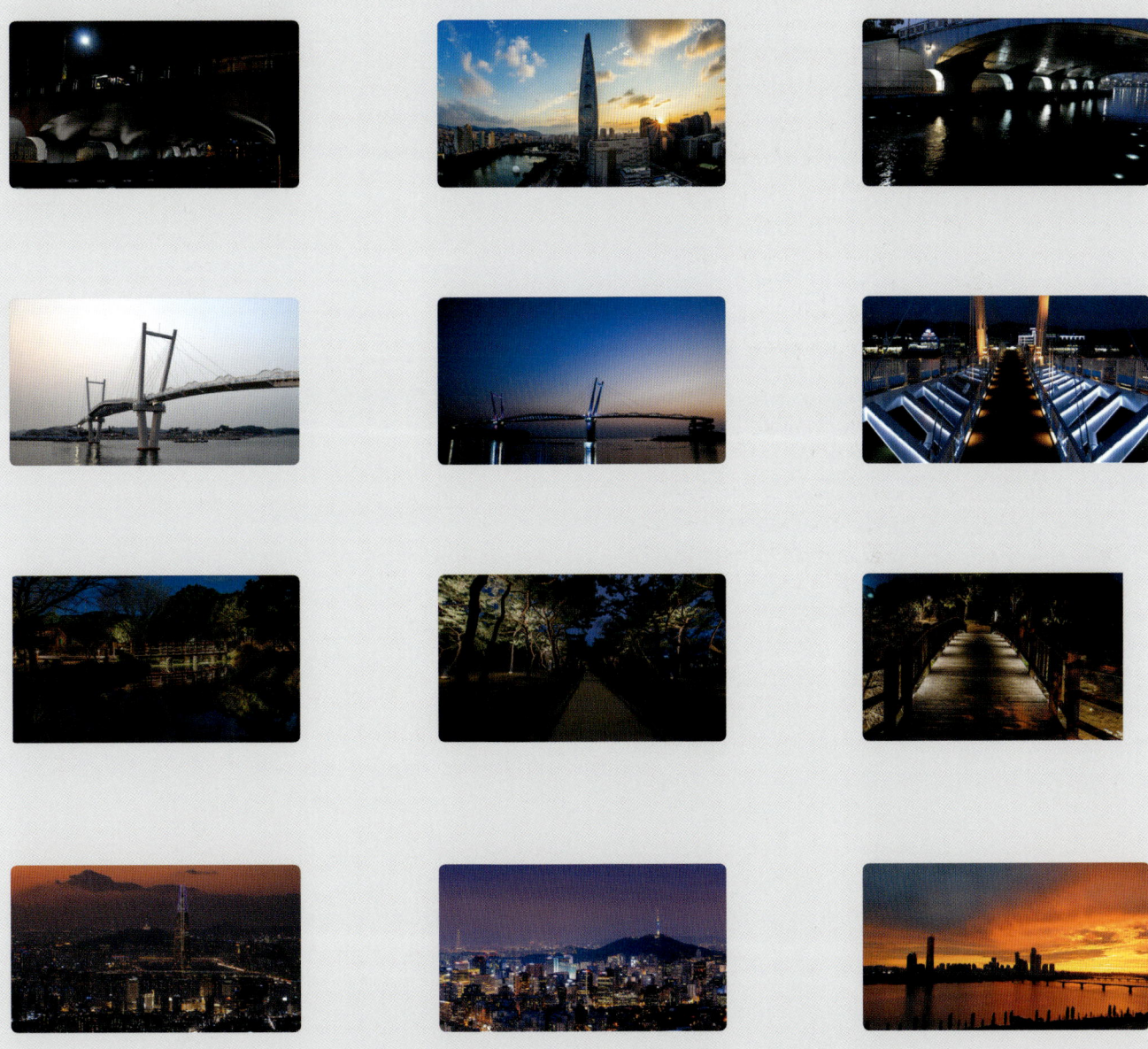

석촌호수,
물의 모습을 빛과 소리로 담다

서울시 잠실 석촌호수 수변 산책로의 빛 연출을 디자인하면서, 호수에 빛을 직접 담았다. 우리는 거기서 낮과 밤의 서로 다른 빛을 보았다. 낮에는 호수 위에 녹조가 있어서 빛이 보이지 않았는데, 밤에는 물속에서 빛이 만들어 내는 작은 원형의 파형을 보았다. 30㎝ 물속에서 50㎝ 깊이의 물속에서 영상연출이 가능한 LED 조명을 디자인하여 설치했다. 그리고 저녁 7시가 되면 숨겨놓은 작은 파이프 스피커에서 물방울 떨어지는 소리와 함께 석촌호수 속의 빛에 물방울이 떨어졌을 때의 아름다운 파형을 호수에 담아냈다.

석촌호수에 빛과 소리를 물에 디자인하면서 빛의 밝기와 움직임, 소리의 울림이 물의 감성을 더 효과적으로 표현해 준다는 사실을 알게 된 빛 연출 작품이었다. 나의 빛은 새로운 소리를 동반한 감성의 언어를 담아냈고, 시민들은 물빛에서 석촌호수의 새로운 모습을 보게 되었다.

SEOKCHON LAKE, CAPTURING THE APPEARANCE OF WATER THROUGH LIGHT AND SOUND

While designing the light production for the waterfront promenade of Seokchon Lake in Jamsil, Seoul, we put the light directly into the lake. During the day, we could not see the light because there was algae on the lake, but at night, we saw small circular waveforms created by the light in the water. We designed and installed an LED light that can project video from 30 centimeters underwater to 50 centimeters deep, and at 7 p.m., a small pipe speaker played the sound of water droplets from a hidden speaker, creating a beautiful waveform in the lake when water droplets fell on the light in Seokchon Lake.

While designing light and sound into the water in Seokchon Lake, I realized that the brightness and movement of the light and the echo of the sound could more effectively express the emotion of the water. My light captured the language of emotion accompanied by new sounds, and the citizens saw a new side of Seokchon Lake in the light of the water.

물결처럼 퍼져나가는 파동을 형상화한 석촌호수 빛의 디자인 | Light Design, visualizing waves spreading like ripples, Seokchon Lake, 2010

서울시 잠실 석촌호수 전경 | Panoramic view of Seokchon Lake, Jamsil, Seoul

서해에서 불어오는 바람을 빛으로 느끼다

FEEL THE BREEZE BLOWING FROM THE WEST SEA THROUGH LIGHT

태안군 안면도에 위치한 해상 인도교인 '대하(大鰕)랑 꽃게랑교'는 보행자 전용 교량이다. 이곳에 서면 탁 트인 서해의 아름다운 바다와 바닷바람을 직접 느낄 수 있다. 일상에서 힘들고 지친 사람들에게 먼바다에서 불어오는 시원한 바람을 맞으며 잠시 쉴 수 있는 공간을 만들어 주고 싶었다.

　　　　우리는 서해의 바람을 빛으로 형상화했다. 어두운 밤에 배와 다리를 스쳐 지나가는 바람을 시각적으로 표현하기 위해 불어오는 바람결에 따라 변화하는 빛을 시각적으로 느낄 수 있도록 빛 연출을 디자인한 것이다. 단순히 빛을 보는 것이 아니라 이곳에서만 느낄 수 있는 분위기를 위한 감성의 빛 연출에 포인트를 뒀다. 시시각각으로 다양하게 변화하는 바람을 부드럽고 따뜻한 색감의 빛으로 시각화하여 사람들에게 행복감을 전해주는 신비로운 공간이 되었다.

'Daeharang Gotgaeranggyo Bridge' is a pedestrian-only footbridge located on Anmyeondo Island, Taeangun. From here, visitors can enjoy the beautiful sea and sea breeze of the open West Sea directly. The aim was to provide a space for people to take a break from their daily lives and relax while feeling the cool breeze from the distant sea.

　　　　To visually represent the wind blowing past the ship and the bridge on a dark night, a light production was designed to change with the wind. The goal was to create an emotional light production for an atmosphere that can only be felt here, not just seen. By visualizing the wind, which changes in different ways at different times, with soft and warm colored light, it has become a mysterious space that delivers a sense of happiness to people.

서해의 바람을 느끼는 태안군 대하(大蝦)랑 꽃게랑교 빛의 디자인 | Light Design, capturing the feeling of the wind from the West Sea, Daeharang and Ggotgaerang Bridge, Taean-gun, 2013

제주도 서귀포 생태공원에 달빛과 별빛을 품다

제주도 서귀포시 하영올래 생태공원의 빛을 디자인하면서, 기존에 설치된 인공적인 조명의 밝기를 낮추고 일정한 시간이 되었을 때 자연의 별빛과 달빛이 만들어내는 서귀포 밤의 자연스러운 모습을 이곳에 담고 싶었다.

나의 빛 연출 콘셉트는 '서귀포의 밤하늘'이었다. 빛을 디자인하는 과정에서 불이 꺼진 맑은 서귀포의 밤하늘 사이로 보이는 별빛과 달빛을 보았고, 그 빛을 공원에 고스란히 담을 수 있도록 빛을 디자인했다. 수림으로 가득한 공간에 작은 빛이 공간을 밝혀준다. 밤 9시가 되면 빛의 밝기는 낮아지고 주변에서 작은 음악이 우리를 아늑하게 감싸 준다. 수림 사이로 감춰져 있던 서귀포 밤하늘에 별이 모습을 드러내는 것이다. 내가 꿈꾼 빛은 아름다운 섬 제주도 서귀포의 밤 자체였다. 나의 빛 연출은 일반적인 조명에서 벗어나 그 장소성을 담아낼 수 있는 자연의 진정성을 담아야 한다고 생각해 왔다. 서귀포 생태공원에서의 빛은 자연 그대로의 밤의 풍광을 살려낸 것이다.

EMBRACING MOONLIGHT AND STARLIGHT, SEOGWIPO ECOLOGICAL PARK, JEJUDO ISLAND

When designing the light for Hayoungolle Ecological Park in Seogwipo City, Jejudo Island, we wanted to reduce the brightness of the existing artificial lights and capture the natural look of the night in Seogwipo City, which is created by natural starlight and moonlight at certain times of the day.

The concept for the light production was 'Seogwipo's night sky'. While designing the light, I saw the starlight and moonlight through the clear, unlit night sky of Seogwipo, and designed the light to reflect that light in the park. The space is filled with trees and small lights illuminate the space. At 9 p.m., the brightness of the light decreases and the ambient music envelops us in a cozy atmosphere. The stars appear in the Seogwipo night sky, hidden among the trees. The light I dreamed of was the night itself on the beautiful island of Seogwipo, Jeju. I have always believed that my light production should be free from general lighting and capture the authenticity of nature that can capture the sense of place. The light at Seogwipo Ecological Park is a natural nighttime landscape.

제주도 서귀포시 걸매생태공원 빛의 디자인 | Light Design, Geolmae Ecological Park ,Seogwipo City, Jeju Island, 2021

서울, 빛의 혈(血)을 발견하다

SEOUL, DISCOVERING THE VEINS OF LIGHT

도시 공간에는 오랜 세월의 무늬가 담긴 고유의 문화가 있다. 빛은 다양한 문화를 효과적으로 담아낸다. 나는 '서울시 야간경관 기본계획' 책임연구원으로 빛을 계획하면서 신비로운 체험을 했다. 바로 대한민국 수도 서울의 힘찬 빛 동맥(動脈)을 본 것이다.

  서울시 당국의 안전 승인을 받아서 서울 도심 한가운데 위치한 남산 N서울타워의 맨 위층의 외부 공간에 올라갔다. 아침부터 저녁까지 변화하는 빛의 시간을 시시각각으로 사진 촬영하면서 '서울의 빛'을 관찰했다. 서쪽 하늘로 해가 서서히 낮아지면서 낮에는 볼 수 없었던, 서울을 밝히는 빛들이 마치 빛의 물결이 다가오듯 빛의 정체를 드러내기 시작했다.

  한강이 밤의 어둠 속으로 잠겨가고, 한강 주변 고속화도로의 자동차에서 흘러나오는 불빛이 퇴근 시간에 맞춰 차츰 밝아왔다. 도로의 빛은 차량이 조금씩 정체되면서 아주 느리게 움직이는 모습을 보였고, 그것은 문득 사람의 혈관처럼 보였다. 그 순간 나는 내 온몸으로 번져오는 전율을 느꼈다. 낮의 회색빛 도시가 밤이 되면 도시의 빛이 생명의 빛으로 변화하여 움직이고, 숨어 있던 보안등과 공원등이 밝게 빛나면서 도시가 마치 하나의 생명체처럼 살아 숨 쉬는 것이었다.

  빛의 맥! 나는 그 순간 도시의 좋은 빛으로 사람들에게 건강한 삶을 선사할 수 있겠다는 생각과 함께, 어둠에 숨겨져 보이지 않는 자연공간에 대해서는 절제된 빛 연출로 '인류가 빚어낸 자연'을 지켜낼 수 있다고 생각했다. 그동안 나는 막연히 도시가 내 꿈을 펼칠 수 있는 무대라고 생각해 왔는데, 서울 지역만의 생동감 넘치는 장엄하고 아름다운 무대를 두 눈으로 확인하게 된 것이다. 그 뒤로 부산시, 대전시, 대구시, 인천시와 울산시 등 여러 도시에서 야간경관 기본계획을 총괄 디자인하면서 그 도시의 생명 줄 같은 빛의 맥을 찾고자 노력했다. 빛이 공간을 아름답게 빚어내는 것도 중요하지만, 철저한 현장 조사를 바탕으로 도시마다 밤의 빛 골격을 찾고, 그 안에 빛의 혈맥이 흐르게 하여 생동감 넘치는 도시의 경관을 창출해 내는 것이 중요했다. 이렇게, 밤의 빛은 살아 숨 쉬어야 한다. 우리가 만들어 낸 빛이 도시의 혈맥이 되어 생동감이 넘치는 밝고 따뜻한 도시를 만들어 가는 것이다.

Urban spaces have their own unique cultures, shaped by patterns that have developed over time. The use of light effectively captures and represents these diverse cultures. While working as the lead researcher for the Master plan of Seoul nightscape, I witnessed the powerful presence of light in Seoul, the capital of Korea, which I saw as a significant and impactful element.

  With the safety approval of the Seoul Metropolitan Government, I climbed to the top floor of Namsan N Seoul Tower, located in the center of Seoul's city center. I observed the 'light of Seoul' by photographing the changing hours of light from morning to evening. As the sun gradually lowered into the western sky, the lights that illuminate Seoul, which were not visible during the day, began to reveal their true colors, as if a wave of light was approaching.

  As the sun set and night fell over the Hangang River, the lights from the cars on the expressways surrounding the river gradually brightened for rush hour. The slow-moving traffic made the lights on the road resemble human blood vessels, sending a chill down my spine. The once gray city transformed into a vibrant, living spectacle at night, with security and park lights illuminating every corner.

  Views of Light! I realized the potential for the city's light to contribute to people's well-being and to protect the natural spaces hidden in darkness. Previously, I had seen the city as a platform for my own dreams, but witnessing Seoul's majestic nocturnal vitality inspired me to design night views in cities like Busan, Daejeon, Daegu, Incheon, and Ulsan. I aimed to capture the essence of each city through thorough field research and use light to create a lively urban landscape. My goal was to ensure that the city's nighttime light was vibrant, alive, and breathed life into the urban environment. The light we design should become the lifeblood of the city, shaping a warm and lively urban environment.

빛의 마스터 플랜, 서울시 야간경관 기본계획 | Light Master Plan of Light Pollution Prevention Plan, Seoul, 2015

빛의 디자인, 시흥시 아브뉴프랑 센트럴 빛의 개선문 | Light Design, Light Triumphal Arch of Avenue France Central, Siheung City, 2021

나에게 빛은 늘 새로운 도전(Challenge)의 대상이었다.
나는 도시와 건축, 그리고 현대 사람들 속에 자리 잡은 빛 환경
문제의 본질을 정확하게 진단하고, 문제를 개선하여 새로운
세계를 창출하는 빛 연출가(Artist of the Light)이다.

Light has always been a new challenge for me.
I am an Artist of light and the General Director of light for
spaces. I have an exceptional ability to diagnose and resolve
light environment issues in cities, architecture, and people,
and I am dedicated to creating new worlds by solving them.

Chapter IV
빛 연출과 공간 마술사: 誠

LIGHT CONDUCTOR AND SPATIAL MAGICIAN: GENERAL LIGHT DIRECTOR

공간은 감성이 담긴 무형의 가치이다.
나는 무형이 지닌 가치를 나의 절제된 빛으로 담아내어
나만의 새로운 공간을 꿈꾼다.

Space is an intangible value that contains emotions.
I dream of a new space of my own by capturing
the intangible value with my understated light and
dream of my own new space.

감성을 담은 빛 연출

나는 공간에 감성의 빛(Soft Light)으로 새로운 생명(Value)을 담는다. 밝고 따뜻한 빛 속에서 삶이 행복해지기를 기원한다. 빛이란 단순히 보이는 것이 아니라, 사람들의 마음 안으로 스며들어야 한다. 여기는 빛 연출가인 나의 이야기와 빛 연출을 꿈꾸는 이들을 위한 자리이다. 빛 연출에 대한 나의 소견, 그리고 빛 연출을 위해 걸어온 길을 소개한다. 이 안에서 빛 연출가의 한 모습을 만나게 되리라 기대한다.

   동서양 문화가 다른 사람도 하루의 일과를 마치고 집으로 향할 때, 서녘 하늘에 불타는 저녁노을 풍경을 만나면 '무엇인가' 가슴 뭉클하게 다가올 때가 있다. 이때의 '무엇인가'가 빛에 대한 감성이 아닐까. 나는 늘 이런 자연 감성의 빛(Soft Light)을 꿈꾼다.

   나는 좋은 빛의 연출이 사람들에게 행복한 순간과 기억을 만들어 줄 수 있다고 생각했다. 좋은 빛 연출에 대한 일념으로 박사과정을 준비하던 시절, 서울시립대학교 배봉관 연구실 좁은 방에 「이연소 조명디자인연구소」를 만들었을 때와 똑같은 열정으로 지금도 감성의 빛을 새롭게 찾고자 변화와 도전 정신으로 빛에 대해 연구를 거듭하고 있다. 빛을 내 손으로 만지고 마음으로 느끼며, 사람들에게 나의 빛 이야기를 들려주는 사람으로 살아간다는 것은 진정 소중한 나만의 행복이다.

   나는 빛 연출가(Artist of the Light)라고 자칭하지만, 나는 빛 자체를 디자인하려고 하지는 않았다. 단지 어둠 속에서 주변을 배려하는 최소한의 빛으로 그 공간만의 감성을 찾아주고 싶었다. 어느 순간부터 빛은 사람들에게 나의 행복을 표현하는 방법이 되어 있었고, 빛은 공간을 이야기하는 나의 또 다른 언어가 되어버렸다. 나의 「좋은 빛 디자인연구소」에서 펼쳐지는 다양한 빛의 테스트를 통해서 나는 무한한 빛을 꿈꾸게 되었다.

   물론 내가 빛으로 새로운 공간을 만들어 가는 동안 힘든 순간도 있었다. 그러나 우리가 믿는 '밝고 따뜻한 빛'은 어둠 속의 나와 우리 스태프에게 우리가 가야 할 길을 밝혀주는 빛이 되었고, 힘이 되어 줬다. 하나의 빛 연출 감성을 찾기 위해 우리는 현장에서 많은 빛과 소리, 영상 테스트를 진행하면서 우리가 찾고자 하는 '그곳만의 장소성'을 만나려고 수 많은 노력을 경주했다. 이는 나를 표현하는 하나의 방법이었고, 자신에 대한 신뢰를 쌓아가는 행동인 동시에 내가 가진 빛에 대한 신념이었다. 그래서 항상 현장에 먼저 간다. 그리고 그곳에서 빛 연출 테스트를 통해 빛을 접하면서 이 공간에 빛이 가지고 있는 매력, 그 빛만이 가지고 있는 행복을 찾으려고 노력했다. 이것은 나를 항상 믿어주는 우리 스태프와 사람들에게 보여주는 나의 진정성이었다.

THE LIGHT PRODUCTION WITH EMOTIONS

I illuminate the space with the gentle warmth of emotion, aiming to bring joy and happiness to those in its glow. Light is not just to be observed, but to be felt within one's heart. This is where I will share my journey as a lighting director and connect with fellow enthusiasts eager to create luminous experiences. Join me as I delve into light production and the path that has brought me here, offering a glimpse into the life of a lighting director.

   When people from different cultures head home after a long day of work, there's something about the sunset in the western sky that touches their hearts. This 'something' may be a feeling for light, and I always dream of this natural soft light.

   I truly believe that creating good lighting can bring happiness and create lasting memories for people. I am as passionate about creating good light now as I was when I was preparing for my Ph.D. program and established the Lee Yeon So Lighting Design Institute in a small room in Baebongkwan's lab at the University of Seoul. I am dedicated to continuing my research on light with a spirit of change and challenge, seeking new ways to create emotional light. Being someone who not only works with light, but also feels it with my heart and shares my light stories with others, brings me genuine happiness.

   I call myself an Artist of the Light, but I did not set out to design the light itself. I just wanted to find the emotion of the space in the darkness, with minimal light that was respectful of its surroundings. At some point, light became a way to express my happiness to people, and light became my other language to talk about space. Through various light tests in my 「Good Light Design Research & Development Center」, I dreamed of infinite light.

   Creating a new space with light had its challenging moments, but the 'bright and warm light' we believed in became a beacon of hope, illuminating our path and instilling strength in me and my team. In our pursuit of a unique lighting approach, we conducted extensive on-site testing with light, sound, and video to capture the essence of the space we seeked. This was not just an expression of myself, but a demonstration of self-assurance and a testament to my faith in the power of light. I always make it a point to visit the site first, seeking to discover how light can enhance the space and bring forth a unique joy through our light production tests. This commitment was my way of staying true to myself and showing gratitude to the individuals who always placed their trust in me.

빛의 디자인, 연세대학교 금호아트홀연세 | Light Design, Kumho Art Hall Yonsei, Yonsei University, 2015

나는 이곳만의 특성과 내용이나 이야기를 잘 모른다. 그래서 나는 빛에 대한 기초 조사와 디자인, 빛 연출 테스트를 통해 내가 알지 못하고 있는 부분, 내가 알려고 하는 부분을 현장에서 찾으려고 노력했다. 결국, 내가 찾으려고 노력했던 나의 빛은 나를 빛 연출가로 가는 바탕이 되어 주었다. 나는 누구보다도 더 뜨거운 열정을 가지고 싶다. 그래서 나는 나 스스로 성장을 꿈꾼다. 그 속에서 새로운 변화를 발견하고 또 다른 새로움에 적극적으로 도전하는 삶을 산다.

I'm unfamiliar with the character, content, or story of this place. Therefore, I conducted on-site research, designed, and performed light production testing in order to fill in the gaps in my knowledge and gain information I desired. Ultimately, my pursuit of knowledge about lighting led me to become a lighting director. I aspire to be more passionate than anyone else and envision a life of continual growth, discovering new changes, and actively embracing new challenges.

빛 연출의 진정한 가치

밤의 어둠을 배려하는 빛

요즘 도시는 너무 밝아지기만 하려 한다. 2012년 한국은 「인공조명에 의한 빛공해 방지법」이 제정되어 관리가 필요할 정도가 됐다. 잘못 설치된 조명은 사람들의 일상생활을 방해하거나 자연환경을 훼손하기도 한다. 나는 이런 문제가 '어둠'을 배려하지 않은 데서 비롯되었다고 본다. 안전한 빛과 아름다운 빛도 중요하지만, 어둠에 대한 배려가 우선 돼야 한다. 무조건 밝게 비춘다고 해서 대상이 잘 보이는 것은 아니다. 빛이 너무 밝으면 오히려 잘 보이지 않는다.

서울을 대표하는 '한강의 밤'을 찾기 위해 한강의 큰 강물 선형을 빛으로 밝혀내기보다는 역발상으로 한강의 밤을 어둡게 연출하여 한강의 윤곽이 뚜렷이 드러나도록 했다. 이는 한강의 자연생태계를 우선시했기 때문이고, 실제로 여의도한강공원에서는 빛의 양적인 팽창보다는 최소한의 빛으로 밤의 풍광을 만들어주기 위해 수변 쪽으로는 오히려 빛의 조도를 낮추는 방법으로 빛을 디자인했다. 빛을 제거하는 방법이 또 하나의 빛 연출이 된 셈이다.

공간의 매력을 담는 빛

경기도 광교 신도시 원천호수공원에서는 매시간, 정시(定時)를 알리는 파스텔 색감의 빛 연출을 통해서 시민들에게 따뜻한 휴식 공간을 제공했다. 공원 본래의 목적대로 시민들에게 자연 쉼터로 이용될 수 있도록 만들었다. 절제된 빛의 연출을 통해서 호수 주변 밤의 모습은 어둠을 배려한 자연의 빛이 되어주었다. 변화된 빛의 칼라는 밤의 시간을 알려주는 원천호수의 매력 있는 빛이 되어 주었다.

THE TRUE VALUE OF LIGHT PRODUCTION

A Light that Cares about the Darkness of the Night

Cities are getting excessively bright in the present day. In 2012, South Korea enacted the Light Pollution Prevention Act to control artificial light pollution. Improperly installed lights can interfere with people's daily lives and even damage the natural environment. I believe this problem stems from a lack of consideration for the darkness. Safe light and beautiful light are important, but consideration for the darkness must be prioritized. Just because you shine brightly does not mean you can see what you're looking at. Too much light can actually make it harder to see.

To find the representative 'Night of the Hangang River' in Seoul, we approached the lighting of the Hangang River differently. Rather than brightening the river with light, we chose to darken it at night to highlight the river's outline. This decision prioritized the natural ecosystem of the Hangang River. Specifically, in Yeouido Hangang Park, we designed the lighting to minimize light on the waterfront side, creating a serene night scenery with minimal lighting. In essence, the reduction of light became a unique way of creating a beautiful night view.

Light to Capture the Charm of a Space

The Woncheon Lake Park in Gwanggyo New City, Gyeonggido, provides a warm and relaxing space for citizens through a pastel-colored light production that announces the hourly time. In addition to the restrained use of light to fulfill the park's original purpose of being used as a natural shelter for citizens and nature, a new space was created with natural light that considers the darkness of the night around the lake. The changed light color has become a charming glow of the source lake that signals the time of night.

빛의 디자인, 소노호텔&리조트 소노캄 여수
Light Design, Sono Hotels & Resorts Sono Calm Yeosu, 2012

국가유산의 가치를 담는 빛

국가유산에 빛을 연출하는 디자인에서는 '국가유산에 대한 보호가 우선'이라는 신념으로 주간의 모습을 생각하고 야간에는 일부 시간에만 국가 문화유산의 특징적인 부분을 중심으로 완전성과 진정성을 나타낼 수 있는 최소한의 빛 연출로 본연의 가치를 담아냈다.

장소성을 담는 빛

빛은 획일화된 조명이 아니라 주변 빛과의 관계를 고려하여 그 공간만의 특성에 따라 빛의 감성과 이야기가 살아있는 종합적인 빛 디자인을 창출해야 한다. 우리가 빛을 디자인한 2012 여수 세계엑스포 5성급 호텔 소노캄(SONO Calm YEOSU)은 여수 밤바다를 배경으로 밤 본연의 어둠을 배려하여 객실에서 흘러나오는 빛과 건축구조 내부의 간접적인 빛을 이용하여 독특한 여수 세계엑스포만의 빛을 창출하는 데 기여했다. 빛 연출을 디자인하면서 이곳에 왜 이 빛이 필요한가? 나의 빛 디자인이 공간과 장소를 훼손시키는 것은 아닐까? 스스로에게 가장 기본적인 질문을 던지고 답을 할 때가 많았다. 나의 빛 연출은 이렇게 마지막까지 기본원칙 지키기에 충실했다.

사람들의 행복한 삶을 밝혀주는 빛

빛 연출은 서로 경쟁하는 것이 아니라, 서로를 배려하는 빛의 위계와 질서를 구축해서 주연과 조연 그리고 효과적인 배경이 디자인될 때 쾌적한 빛이 연출될 수 있다. 빛 연출이란 빛 자체가 중요한 것이 아니라 빛이 만들어 낸 감성의 공간에서 사람들이 행복을 느낄 때 진정한 빛의 가치가 살아난다. 따라서 모든 빛의 시작과 끝은 사람에게 행복을 선사하기 위해 존재한다.

The Light Capturing the Value of National Cultural Heritage

In the light design for the scenery of the national cultural heritage, we thought of the daytime appearance with the belief that 'the protection of the national cultural heritage comes first', and at night, we focused on the characteristic parts of the national cultural heritage only during some hours, and created a minimal light production to show the integrity and authenticity of the site.

Light that Captures a Sense of Place

Lighting should not just be uniform illumination, it should be a thoughtful design that takes into account the surrounding light and creates an emotional and narrative impact based on the space's characteristics. SONO Calm YEOSU, a five-star hotel for the 2012 Yeosu World Expo, where we designed the lighting, contributed to creating a unique landscape by considering the darkness of the night against the backdrop of the Yeosu night sea, using light from the guest rooms and indirect light from inside the architectural structure. When designing lighting, I ask myself basic questions: why is light needed here? Will my design detract from the space? I always ask and answer these fundamental questions, remaining true to this principle throughout the design process.

Light that Brightens People's Happy Lives

Light production does not compete with each other, but builds a hierarchy and order of light that considers each other, so that a pleasant light can be created when the main characters, supporting characters, and effective background are designed. The true value of light comes to life when people feel happy in the emotional space created by the light, not the light itself. Therefore, the beginning and end of all light exists to bring happiness to people.

빛의 디자인, 삼척시 죽서루(竹西樓) | Light Design, Jukseoru, Samcheok City, Gangwon-do, 2018

푸른 바다,
그 깊음이 나를 부르고
배움은 끝없이 새로운 물결로 나에게 다가온다.

The blue sea,
Its depth calls to me,
And learning endlessly approaches me like new waves.

빛의 디자인, 미디어 아트 빛의 바다, 속초 | Light Design, Sea of Lights, Sokcho, 2025

속초 해변에 담은 미디어 아트, 빛의 디자인
파란 바다와 금빛 모래가 만나는 속초 해수욕장.
우리는 동해의 물결을 빛으로 담아내어,
끝없이 펼쳐진 자연의 빛을 디자인했다.
한때 여름만의 공간이었던 이곳이,
이제 사계절 내내 다채로운 빛으로 새롭게 피어난다.

Media art captured at Sokcho Beach, the design of light.
Sokcho Beach, where the blue sea meets the golden sand.
We captured the waves of the East Sea in light and designed the endless natural light that stretches out.
This place, once a space only for summer, is now blooming anew with vibrant lights throughout all four seasons.

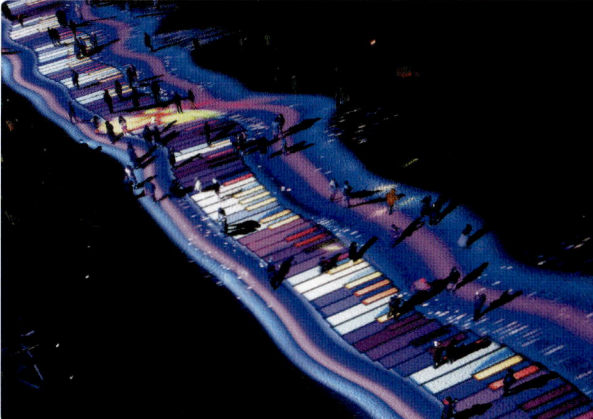

빛 연출가가 되려는 이에게

빛은 나에게는 생명이다. 사람과 사람을 이어주는 따뜻한 마음이 내가 꿈꾸는 빛이다. 내가 빛 연출가가 되려는 작은 마음으로 빛을 꿈꾸던 시절부터 나는 이성의 빛이 아닌 사람의 마음을 담는 감성의 빛을 먼저 꿈꿨다. 이는 서울시립대학교 배봉관 연구실 한 모퉁이에 「이연소 조명디자인연구소」를 만들 때부터 가진 마음이었다. 그즈음부터 나는 감성의 빛을 찾으려는 빛 연구를 지속해 왔고, 지금까지 나는 사람들의 행복을 위해서 따뜻한 나의 빛과 함께 더불어 살아가고 있다.

나는 빛 연출 총감독으로서 내가 이루고자 하는 빛의 세계를 꿈꾼다. 나에게 도시와 공간은 비어 있는 하얀 캔버스(Canvas)와 같다. 나는 빛이라는 붓과 물감으로 도시와 공간을 밤 속에서 최소한의 빛으로 물들여 나간다. 우리의 무대는 도시와 이곳에 숨어 있는 소중한 가치를 담고 있는 옛 궁궐이나 공원, 마을, 서로를 이어주는 다리, 즐거움이 함께하는 건축공간 등, 사람이 행복하게 살아가는 삶의 가치가 담겨있는 소중한 곳들이다. 특히 도시 속에서 국가 문화유산에 대한 빛 연출은 극도의 긴장감이 요구된다. 문화유산 본연의 풍광을 훼손시키지 않는 범위에서, 빛의 분위기와 연출 방법을 찾아내는 것이 중요하기 때문이다.

나는 빛을 디자인하려고 하기보다는 이곳만의 장소성을 찾고자 노력한다. 어둠 속에서 주변을 배려하는 절제된 최소한의 빛이 공간의 장소성을 담아낼 수 있기만을 바랄 뿐이다. 이러한 빛을 찾아내고 생명처럼 밝혀지기까지는 많은 스태프와 함께 공동작업을 진행한다. 현장에서 다양한 빛 연출 테스트를 통해 빛을 접하며 빛이 그 공간에서 지닌 매력, 빛의 가치를 찾고자 했다. 이 과정에서 수많은 스태프와 함께 고민하고 함께 그곳의 가치를 담고자 노력한 결과 속에서 우리만의 빛을 만들어 낼 수 있었다. 빛 연출은 그만큼 다양한 장르가 함께 만들어가는 종합예술이다.

1) Value, 빛의 가치를 지키려는 마음

빛 연출가는 가장 먼저 겸손한 마음을 가져야 한다. 공간과 대상에 자신의 마음을 담아 새로운 작품을 만들어 내고, 작품을 통해 사람들에게 빛의 진정성이 전달되어야 하는 것이다. 달리 말하면, 빛으로 공간과 대상에 조명을 입히는 것이 아닌, 그 장소가 지닌 사연을 빛의 정서로 느끼도록 해야 한다. 이러한 빛 연출가는 늘 자신이 추구하는 순수한 가치를 지향하면서 겸허한 마음으로 배우려고 할 때 새로운 빛을 세상에 만들어 낼 수 있다.

FOR INDIVIDUALS WHO ASPIRE TO BECOME A LIGHTING DIRECTOR

Light is life to me. The warm heart that connects people to people is the light I dream of. From the time I dreamed of light with a small heart to become a light director, I first dreamed of the light of emotion, not the light of reason, but the light of people's hearts. This was the mindset I had from the time I created the Lee Yeon So Lighting Design Institute in a corner of Baebongkwan's lab at University of Seoul. Since then, I have been researching light to find the light of emotion, and until now, I have been living together with my warm light for the happiness of people.

As a light director, I dream about the world of light I want to create. For me, cities and spaces are like blank white canvases. With the brush and paint of light, I color the city and space with minimal light in the night. Our stage is the city and the precious values that are hidden in it, such as old palaces, parks, villages, bridges that connect people to each other, and architectural spaces that bring joy to people's lives. Especially in cities, the light production of national cultural heritage requires extreme tension. It is important to find the right mood and way to play with light without destroying the original scenery of the heritage site.

I try to find a sense of place rather than trying to design light. I hope that an understated, minimal light that is respectful of its surroundings in the darkness can capture the sense of place of a space. It takes a lot of collaboration with my staff to find this light and bring it to life. Through various light production tests in the field, we encountered light and tried to find the charm and value of light in the space. In this process, we were able to create our own light as a result of thinking with many staff members and trying to capture the value of the space together. Light production is a multidisciplinary art that involves many different genres.

1) Value, a Heart for the Value of Light

The first thing a light director must have is a humble heart. They should create new works by putting their heart into spaces and objects, and convey the authenticity of light to people through their works. In other words, they should not just illuminate spaces and objects with light, but make people feel the story of the place with the emotion of light. Such a light director can create new light in the world when he or she is willing to learn with a humble heart, always aiming for the pure values he or she pursues.

2) Challenge, 변화와 도전을 꿈꾸는 마음

빛 연출가는 혼자가 아닌, 종합예술가의 마음처럼 함께 만드는 배려의 마음을 가져야 한다. 빛 연출은 빛과 공간을 새롭게 디자인하지만, 문학, 음악, 연극, 영화, 뮤지컬, 춤과 노래 등 다양한 예술을 비롯하여 미디어아트, 컴퓨터 그래픽 등 다양한 문화 콘텐츠를 포함하고 있어서 종합적인 예술과도 같다. 빛의 디자이너를 꿈꾸는 사람이라면 당장 눈앞에 단편적인 현상을 보기보다는 다양한 견문을 쌓으라고 권장하고 싶다. 많은 곳을 여행하고, 많은 것을 보고 느끼며, 많은 것을 경험하면서 많은 사람 속에서 새로움과 변화를 꿈꾸라는 것이다. 빛 연출가의 사고는 늘 깨어 미래를 향해 열려 있어야 하고 시대의 유행이나 변화의 물결을 온몸으로 받아들여야 한다. 빛을 꿈꾸는 사람은 감성뿐만 아니라 나날이 발전하는 전기, 전자 분야의 공학과 AI, 다양한 예술과 함께 호흡해야 한다. 빛과 공간의 디자이너이자 연출가, 총감독은 산업 전사이자 예술가여야 한다는 말을 다시 상기할 필요가 있다.

3) Mind, 세상을 이롭게 하려는 따뜻한 마음

나는 디자인 대학에서 강의할 때, 학생들을 가르치면서 직업으로 빛 연출가를 선택하라는 말을 절대 하지 않았다. 디자인을 전공한 사람이 분야가 전혀 다른 분야의 사업가가 될 수도 있고, 맛있는 요리를 만드는 요리사, 더 좋은 세상을 꿈꾸는 사람이 될 수도 있기 때문이다. 혹은 전혀 다른 분야의 전문가가 될 수도 있다. 사람의 앞날은 그 누구도 알 수가 없다. 이는 미리 자신의 숨어 있는 달란트를 예단하지 않으며, 일정한 울타리를 미리 치지 않으면서 늘 새롭게 열린 마음으로 세상을 바라보는 사람이 되라는 것이다. 자신의 마음이 향하는 곳이 사람을 위한 마음이 되고, 내가 꿈꾸는 빛이 세상을 이롭게 할 수 있는 빛이 된다고 생각한다면 훌륭한 빛의 디자이너로 성장할 수 있다고 생각한다.

2) Challenge, a Heart that Dreams of Change

A light director must have a caring heart that creates together, not alone, like the heart of a multidisciplinary artist. Light design involves redesigning light and space, but it is also a comprehensive art that includes various arts such as literature, music, theater, film, musicals, dance and song, as well as various cultural contents such as media arts and computer graphics. If you are an aspiring light designer, I would encourage you to travel to many places, see many things, feel many things, experience many things, and dream of newness and change in many people. A light director's mind should always be open to the future and embrace the trends and waves of change with their whole body. A person who dreams of light is not only an emotional person, but also a person who breathes together with the ever-advancing engineering of electricity and electronics, AI, and various arts. It is worth recalling that the designer, director, and general manager of light and space must be an industrial warrior and an artist.

3) Mind, a Warm Heart that Wants to Do Good in the World

When I taught at a design university, I never told my students to choose a career as a light director. A person who majored in design can become a businessman in a completely different field, a chef who creates delicious dishes, a person who dreams of a better world, or an expert in a completely different field. You never know what your future holds. This is why it is important to be a person who looks at the world with a fresh and open mind, without prejudging one's hidden talents, and without putting up certain fences. If you think that the place where your heart is directed becomes a heart for people, and the light you dream about becomes a light that can benefit the world, I think you can grow into a great light designer.

빛 연출가
이연소가 걸어온 길

나는 대학에서 미술을 전공했지만 도시와 건축, 국가유산인 궁궐과 종묘, 그리고 음악과 그림, 연극, 영화, 뮤지컬 등 다양한 장르의 작품을 찾아다니면서 젊은 시절을 보냈다. 그 뒤, 한국 고 건축(古建築) 미학(美學)을 접하면서 「문화재 조명에 대한 연구」를 통해 한국 고 건축 공간 속에서 건축물과 빛이 만들어 내는 새로운 공간 미학을 연구하게 됐다. 서울시립대학교 조경학과 박사과정에 입학해서 빛과 경관의 연구(「문화재 야간경관에 미치는 조명 물리량 연구」, 박사학위논문, 2012)를 본격적으로 시작했다.

　　　　한국적인 미학을 빛으로 찾고자 했던 나는 자신만의 이야기를 빛으로 나타낼 수 있는 '절제의 빛, 감성의 빛(Soft Light)'이라는 빛 연출 언어를 만들어 냈다. 나의 이런 연구와 새로운 도전은 야간경관에 대한 개념이 없던 시절, 국가 문화유산 경관조명과 도시 공간의 조명계획을 만들어 내는 도시 야간경관 기본계획을 최초로 시작하는 계기가 됐다.

　　　　나는 이 분야 전문가로 서울시, 부산시, 인천시, 대전시, 대구시, 울산시 등의 야간경관 기본계획 책임연구원으로 도시의 빛 마스터플랜을 새롭게 제시했다.

　　　　나는 공간을 밝히는 기존의 조명 관점에서 벗어나 '감성의 빛 연출' 개념을 도입하여 청계천 3공구 빛의 디자인 설계, 서울 한강르네상스 빛의 마스터플랜, 서울시 한강교량과 11개 한강공원 빛의 디자인과 설계, 경복궁, 창덕궁, 창경궁의 야간경관 조명설계와 함께 '2013 창덕궁 달빛기행', '2024 창경궁 물빛연화', '2025 미디어아트 빛의 바다, 속초' 등의 빛 연출 설계를 총괄 진행했다.

　　　　현재 세계조명디자이너협회(IALD) 정회원, (사)한국경관학회 부회장, (사)이코모스 한국위원회 정회원, 서울시 좋은빛위원회 위원, 경기도와 강원도, 대구광역시 경관위원회 위원, 인천광역시 공공디자인위원회 위원으로 활동하고 있다.

　　　　2006년 서울시립대학교 배봉관 연구실에서 「이연소조명디자인연구소」라는 간판을 내걸고 기존의 밝게만 만드는 볼거리 중심의 경관조명 설계에서 "밤 자체를 배려한다."라는 빛의 새로운 관점을 제시한 빛의 디자이너이자 빛 연출가이다.

　　　　현재 ㈜유엘피(www.ulplighting.com)의 설립자로서, ㈜유엘피 총감독으로 활동 중이다.

THE PATH OF LIGHT
DIRECTOR LEE YEON SO

I majored in art in college and spent my youth exploring cities, architecture, and various art forms like music, painting, theater, and movies. My interest in Korean ancient architecture led me to study the aesthetics of light in these spaces. I pursued a Ph.D. in Landscape Architecture at the University of Seoul, researching lighting for cultural properties ("A Study on the Physical Quantity of Lighting on Cultural Property Night Scenery," 2012). I developed a lighting language called 'Soft Light' to express personal stories through light.

　　　I initiated the first urban night landscape master plan, creating lighting designs for national heritage and urban spaces. As a specialist in this field, I led night view master plans for cities like Seoul, Busan, and Incheon. I introduced the concept of 'emotional light production' and designed lighting for significant projects like Cheonggyecheon District, the Hangang Renaissance Plan, Seoul's Hangang bridges, Gyeongbokgung Palace, and Changdeokgung Palace.

　　　Currently, I'm a full member of the International Association of Lighting Designers (IALD), Vice President of the Korea Landscape Council (KLC), and serve on multiple committees related to landscape and public design across several regions in Korea.

　　　I am a light designer and light director who presented a new perspective of light, "Consider the night itself," from the existing brightness-oriented landscape lighting design to "consider the night itself," at the Baebongkwan Laboratory of University of Seoul in 2006.

　　　I am the founder of ULP (www.ulplighting.com) and currently serve as the general director of ULP Co. Ltd,.

국내 수상 경력
- 제1회 서울특별시 좋은빛상, 서울시 한강 야간경관 기본계획 및 빛의 디자인, 2012.
- 제10회 대한민국 조경대상, 문화재청장상, 창덕궁 달빛기행 빛의 디자인, 2019.
- 제18회 한국색채대상, 대상, 국토교통부장관상, 인천시 10대 야경명소 빛의 디자인, 2020.

국제 수상 경력
- 국제도시조명연맹(LUCI), CPL「City.People.Light」 Award, First prize, 2008 LUCI AGM : 서울시 청계천 복원건설공사 빛의 디자인 1등, 2008
- 국제도시조명연맹(LUCI), CPL「City.People.Light」 Award, Second prize, 2013 LUCI AGM : 서울시 여의도한강공원 빛의 디자인 2등, 2013
- 국제도시조명연맹(LUCI), CPL「City.People.Light」 Award, Second prize, 2016 LUCI AGM : 서울시 경의선 숲길 빛의 디자인 2등, 2016

관련 위원회 활동
- 서울특별시 좋은빛위원회 위원
- 강원특별자치도 경관위원회 위원
- 경기도 경관위원회 위원
- 대구광역시 경관위원회 위원
- 인천광역시 공공디자인위원회 위원

국내 소속단체
- (사) 한국경관학회 부회장
- (사) 한국전통조경학회 상임이사
- (사) 한국환경조명학회 이사
- (사) 한국미디어아트협회 정회원
- (사) 한국조명디자이너협회 정회원

국제 소속단체
- 이코모스 정회원
- 국제조명디자이너협회 정회원

Domestic Awards
- 1st Seoul Good Light Award, Basic Plan and Light Design for Night View of Hangang River, Seoul, Korea, 2012.
- 10th Korea Landscape Award, Grand Prize, Minister of Cultural Heritage Administration, Changdeokgung Palace Moonlight Path Light Design, 2019.
- 18th Korea Color Award, Grand Prize, Minister of Land, Infrastructure, and Transport Award, Light Design for Incheon's Top 10 Night Scenic Spots, 2020.

International Awards
- Lighting Urban Community International (LUCI), CPL「City.People.Light」 Award, First prize, 2008 LUCI AGM : 1st prize for light design for Cheonggyecheon Restoration and Construction Project, Seoul, Korea, 2008
- Lighting Urban Community International (LUCI), CPL「City.People.Light」 Award, Second prize, 2013 LUCI AGM : 2nd prize, Yeouido Hangang Park Light Design, Seoul, Korea, 2013
- Lighting Urban Community International (LUCI), CPL「City.People.Light」 Award, Second prize, 2016 LUCI AGM : 2nd place, Gyeonguiseon, Railway Track, Forest Road Light Design, Seoul, Korea, 2016

Committee Activities
- Member of Seoul Metropolitan Government Good Light Committee
- Member of Gangwon State Landscape Committee
- Member of Gyeonggi Province Landscape Committee
- Member of Daegu Metropolitan City Landscape Committee
- Member of Incheon Metropolitan City Public Design Committee

Domestic Affiliations
- Vice President, Korea Landscape Council (KLC)
- Executive Director, Korean Institute of Traditional Landscape Architecture (KITLA)
- Director, Korea Society of Lighting and Visual Environment (KSLVE)
- KMAA member (Korea Media Art Association)
- KALD member (Korea Association Lighting Designers)

International Affiliations
- ICOMOS member (International Council on Monuments and Sites)
- IALD member (International Association of Lighting Designers)

REFERENCES

단행본 및 논문

- 이연소, 「도시 내의 문화재 야간조명연출에 관한 연구」, 명지대학교 건축공학과 석사학위논문, 2000
- 이연소, 「문화재 야간경관에 미치는 조명(照明)물리량 연구」, 서울시립대학교 조경학과 박사학위논문, 2012
- 이연소, 「역사문화도시의 수변 경관 조명(照明) 개선방안」, 『한국전통조경학회』, 2011
- 최기수 외, 「오늘, 옛 경관을 다시 읽다 : 전통조경에 담겨 있는 선조들의 멋과 지혜」, 파주: 도서출판 조경, 2007(확인 요망)
- 프롬나드디자인연구원, 『프롬나드 디자인: 디자인의 미래, 디자인 정책을 생각하며』, 서울: 한국학술정보, 2009

보고서 및 웹사이트

- 강원도 원주시 '나오라쇼' 빛의 디자인, 원주시, 2021
- 강원도 원주시 '소금산 스카이벨리' 빛의 디자인, 원주시, 2021
- 경기도 빛 공해 환경영향평가 및 측정·조사, 경기도, 2014
- 경복궁 별빛야행 빛의 디자인, 문화재청, 2017
- 광교신도시 호수공원 빛의 디자인, 광교신도시, 2009
- 구미시 야간경관 기본계획, 구미시, 2012
- 김해 롯데워터파크 빛의 디자인, 2015
- 당진시 야간경관 기본계획, 당진시, 2021
- 대구광역시 야간경관 기본계획, 대구광역시, 2010
- 대구광역시 야간경관 기본계획, 대구광역시, 2019
- 대전광역시 야간경관 기본계획, 대전광역시, 2010
- 대전광역시 야간경관 재정비계획, 대전광역시, 2018
- 문경새재 빛의 디자인, 문경시. 2015
- 보물 제1호 서울 흥인지문 빛의 디자인, 종로구청, 2015
- 부산광역시 북항재개발사업 공원과 8개 교량 빛의 디자인, 부산항만공사, 2016
- 부산광역시 야간경관 기본계획, 부산광역시, 2014
- 사적 제10호 서울성곽 빛의 디자인, 서울특별시, 2005
- 사적 제280호 한국은행 화폐금융박물관 빛의 디자인, 한국은행, 2005
- 서울시 경의선 숲길 빛의 디자인, 서울특별시, 2016
- 서울시 빛 공해 방지계획, 서울특별시, 2014
- 서울시 빛 공해환경영향평가 및 측정·조사, 서울특별시, 2013
- 서울특별시 야간경관 기본계획, 서울특별시, 2015
- 서울특별시 야간경관 기본계획. 서울특별시, 2008
- 안산시 야간경관 기본계획, 안산시, 2017
- 여주시 신륵사 출렁다리 빛의 디자인, 여주시, 2021
- 연세대학교 금호아트홀 빛의 디자인, 금호건설, 2012
- 울산광역시 빛 공해환경영향평가 측정·조사, 울산광역시, 2016
- 울산광역시 야간경관 기본계획, 울산광역시, 2015
- 원주시 야간경관 기본계획, 원주시, 2008
- 원주시 야간경관 기본계획, 원주시, 2014
- 인천광역시 야간경관 기본계획, 인천광역시, 2018
- 인천국제공항 아시아나항공 라운지 빛의 디자인, 아시아나항공, 2016
- 죽녹원 빛의 디자인, 담양군, 2014
- 창덕궁 달빛기행 빛의 디자인, 창덕궁관리사무소, 2015
- 청계천 복원건설공사 3공구 빛의 디자인, 서울특별시, 2004
- 청주시 야간경관 기본계획, 청주시, 2010
- 충주시 야간경관 기본계획, 충주시, 2012
- 태안군 드르니항 빛의 디자인 2013
- 한강공원지구 및 연변 민간 건축물 경관조명 기본설계, 서울특별시, 2008
- 한강변 경관조명 기본 및 실시설계, 서울특별시, 2007
- www.cha.go.kr
- www.heritage.go.kr
- www.pexels.com
- www.pixabay.com
- www.seoul.go.kr
- www.standard.go.kr
- www.sonohotelsresorts.com
- www.ulplighting.com
- https://hangang.seoul.go.kr
- www.wikipedia.org
- www.lak.co.kr

Books and Articles

- Lee, Yeon So, "A Study on Night Lighting of Cultural Properties in Urban Areas," Master's Thesis, Department of Architecture, Myeongji University, 2000
- Lee, Yeon So, "A Study on the Physical Quantity of Lighting on Cultural Property Night Scenery," Ph.D. Thesis, Department of Landscape Architecture, Seoul National University, 2012
- Lee, Yeon So, 「Improvement Plan for Waterfront Landscape Lighting in Historical and Cultural Cities」, Korean Society of Traditional Landscape Architecture, 2011
- Lighting and Interiors Editorial Department, Lighting & Interiors Lighting & Interiors, Join Media Group, 2022
- Choi, Key Soo, et al, 「Today, Rereading the Old Landscape : The Style and Wisdom of the Ancestors in Traditional Landscapes」, Paju: Book Publishing Landscape, 2007
- Promenade Design Institute , 『Promenade Design: Thinking about the Future of Design, Design Policy, Seoul: Korea Academic Information, 2009

Reports

- 'Naora Show' Light Design in Wonju, Gangwon-do, Korea, 2021
- Light Design for 'Sogeumsan Mountain Sky Valley', Wonju-si, Gangwon-do, Korea, 2021
- Gyeonggi-do Light Pollution Environmental Impact Assessment and Measurement and Survey, Gyeonggi-do, 2014
- Gyeongbokgung Palace Starry Night Light Design, Cultural Heritage Administration, 2017
- Gwanggyo New City Lake Park Light Design, Gwanggyo New City, 2009
- Gumi City Night View Basic Plan, Gumi City, 2012
- Gimhae Lotte Water Park Light Design, 2015
- Dangjin City Night Scene Basic Plan, Dangjin City, 2021
- Daegu Metropolitan City Night View Basic Plan, Daegu Metropolitan City, 2010
- Daegu Metropolitan City Night View Basic Plan, Daegu Metropolitan City, 2019
- Daejeon Metropolitan City Night View Basic Plan, Daejeon Metropolitan City, 2010
- Daejeon Metropolitan City Night Scene Reorganization Plan, Daejeon Metropolitan City, 2018
- Mungyeong Saejae Light Design, Mungyeong City, 2015
- Treasure No. 1 Seoul Heunginjimun Gate Light Design, Jongno-gu Office, 2015
- Light Design for Bukhanjae Redevelopment Project Park and 8 Bridges, Busan Port Authority, 2016
- Basic Plan for Night Scenery in Busan, Busan Metropolitan City, 2014
- Historic Site No. 10 Seoul Castle Light Design, Seoul Metropolitan Government, 2005
- Historic Site No. 280 Light Design for the Bank of Korea Museum of Money and Finance, Bank of Korea, 2005
- Light Design for Gyeongui-Seon Forest Path, Seoul, Korea, 2016
- Light Pollution Prevention Plan for Seoul, Seoul, 2014
- Environmental Impact Assessment and Measurement and Survey of Light Pollution in Seoul, Seoul, 2013
- Basic Plan for Night Scenery in Seoul, Seoul, 2015
- Basic Plan for Night Scenery in Seoul. Seoul, 2008
- Ansan City Night View Basic Plan, Ansan City, 2017
- Light Design for Chulungdari Bridge, Shillaeksa Temple, Yeoju-si, 2021
- Light Design for Kumho Art Hall, Yonsei University, Kumho E&C, 2012
- Ulsan Metropolitan City Light Pollution Environmental Impact Assessment Measurement and Survey, Ulsan Metropolitan City, 2016
- Ulsan Metropolitan City Night View Basic Plan, Ulsan Metropolitan City, 2015
- Wonju Night Scenery Basic Plan, Wonju, 2008
- Wonju Night View Basic Plan, Wonju, 2014
- Incheon Metropolitan City Night View Basic Plan, Incheon Metropolitan City, 2018
- Light Design for Asiana Airlines Lounge, Incheon International Airport, Asiana Airlines, 2016
- Juknokwon Light Design, Damyang-gun, 2014
- Changdeokgung Palace Moonlight Procession Light Design, Changdeokgung Palace Management Office, 2015
- Light Design for Cheonggyecheon Restoration Project 3, Seoul, Korea, 2004
- Cheongju City Night View Basic Plan, Cheongju City, 2010
- Chungju Night Scenery Basic Plan, Chungju, Korea, 2012
- Light Design for Drni Port, Taean-gun, Korea, 2013
- Basic Design of Landscape Lighting for Hangang Park District and Private Buildings along the River, Seoul, 2008
- Basic and Implementation Design for Landscape Lighting along the Hangang River, Seoul, 2007

PHOTO CREDIT

Most of the images of the artworks were provided by ULP.
The sources of the other images used are as follows.

- 002-003 https://pixabay.com/photos/pantheon-rome-italy-black-and-white-5010757/
- 016-017 https://pixabay.com/ko/photos/%EB%8F%84%EC%8B%9C-%EA%B1%B4%EB%AC%BC-%EC%95%BC%EA%B2%BD-%EC%84%9C%EC%9A%B8-7459162/
- 020-021 https://pxhere.com/ko/photo/1225110
- 024-025 https://pixabay.com/ko/photos/%EB%91%90%EB%A C%BC%EB%A8%B8%EB%A6%AC-%EC%97%AC%EB%AA%85-%EC%83%88%EB%B2%BD-%EC%9D%BC%EC%B6%9C-3559321/
- 027 https://www.pexels.com/ko-kr/photo/127577/
- 029 https://pxhere.com/ko/photo/928257
  https://pixabay.com/photos/beach-sea-sand-thailand-4852830/
- 036 https://pixabay.com/photos/seoul-cheonggyecheon-stream-3804293/
- 038-039 https://pxhere.com/ko/photo/842012
- 042-043 https://pixabay.com/ko/photos/%EB%B6%80%EC%82%B0-%EB%B0%A4-%EC%A0%80%EB%85%81-%EB%8F%84%EC%8B%9C-%EB%8F%84%EC%8B%9C%EC%9D%98-7241479/
- 045 https://pixabay.com/ko/photos/%EB%8F%84%EC%8B%9C-%EC%84%9C%EC%9A%B8-%EC%9D%BC%EB%AA%B0-%ED%92%8D%EA%B2%BD-7111378/
- 050 https://pixabay.com/photos/nature-grass-field-rural-4607496/
  https://pxhere.com/en/photo/1369330
  https://pxhere.com/en/photo/745871
- 051 https://pixabay.com/hu/photos/hull%C3%A1mok-tenger-%C3%B3ce%C3%A1n-strand-3473335/
  https://pixabay.com/hu/photos/ter%c3%bclet%c3%a9n-f%c3%b6ld-felh%c5%91k-%c3%a9g-horizont-533541/
  https://pixabay.com/fi/photos/vesist%C3%B6-costa-ranta-matkailu-meri-3045871/
- 052 https://pixabay.com/el/photos/%CE%B9%CF%83%CE%BB%CE%B1%CE%BD%CE%B4%CE%AF%CE%B1-aurora-borealis-2111811/
  https://www.pickpik.com/aurora-borealis-northern-lights-forest-trees-woods-silhouettes-37935
  https://pixabay.com/es/photos/finlandia-northern-lights-2638253/
- 053 https://pixabay.com/vi/photos/%c3%a1nh-s%c3%a1ng-%c4%91%c3%aam-nh%c3%a0-th%e1%bb%9d-t%e1%bb%91i-n%e1%ba%bfn-nh%c3%a0-1568056/

https://pixabay.com/cs/photos/pry%C4%8D-sv%C4%9Btlo-lesn%C3%AD-cesta-3294679/
https://pixabay.com/bg/photos/%D0%B4%D1%8A%D1%80%D0%B2%D0%B5%D1%82%D0%B0-%D0%B3%D0%BE%D1%80%D0%B0-%D0%B7%D0%B2%D0%B5%D0%B7%D0%B4%D0%B8-%D0%BC%D0%B8%D1%81%D1%82%D0%B8%D1%87%D0%B5%D0%BD-5159688/

- 054 https://pixabay.com/ja/photos/%E3%83%A9%E3%82%A4%E3%83%88%E3%83%8B%E3%83%B3%E3%82%B0-%E9%9B%B7-%E9%9B%B7%E9%9B%A8-%E5%B5%90-1056419/
  https://pixabay.com/id/photos/matahari-terbenam-senja-petir-badai-2530165/
  https://www.wallpaperflare.com/cityscape-with-thunder-california-storm-weather-thunderstorm-wallpaper-ukarp
- 055 https://pixabay.com/de/photos/nebel-natur-drau%C3%9Fen-b%C3%A4ume-wald-3902368/
  https://pixabay.com/ko/photos/%ED%83%9C%EC%96%91%EC%97%B4-%EC%88%B2-%EC%95%88%EA%B0%9C-%EB%82%98%EB%AC%B4-1547273/
  https://pixabay.com/es/photos/naturaleza-vista-paz-pastar-3362956/
- 056 https://commons.wikimedia.org/wiki/File:Tokyo,_Japan_(43696647564).jpg
- 058 https://pixabay.com/ko/photos/%EA%B1%B4%EC%B6%95%EB%AC%BC-%EB%A5%B4-%EC%BD%94%EB%A5%B4%EB%B7%94%EC%A7%80%EC%97%90-%EB%A1%A0%EC%83%B9-827805/
- 059 https://pixabay.com/photos/notre-dame-du-haut-de-ronchamp-372585/
- 060-061 https://pixabay.com/photos/mountains-fog-fall-landscape-6964950/
- 074-075 https://pixabay.com/photos/hong-kong-cityscape-night-city-3788013/
- 076-077 https://pxhere.com/ko/photo/1612216
- 080 https://pixabay.com/photos/opera-house-sydney-australia-4670527/
- 081 https://pixabay.com/photos/sydney-opera-house-sydney-australia-361698/
  https://pixabay.com/photos/sydney-opera-house-building-363244/
  https://pixabay.com/photos/sydney-australia-night-city-792276/

- **082-083** https://pixabay.com/ko/photos/%EB%8F%84%EC%8B%9C-%EA%B1%B4%EB%AC%BC-%EC%95%BC%EA%B2%BD-%EC%84%9C%EC%9A%B8-7459162/
- **086** https://pixabay.com/photos/seoul-city-skyscraper-river-410269/
- **087** https://pixabay.com/photos/sunset-han-river-han-river-park-7467688/
- **091** https://pixabay.com/photos/han-river-south-korea-city-7051295/
- **092-093** https://pixabay.com/photos/south-korea-south-korea-han-river-2487786/
- **094-095** https://pixabay.com/photos/night-view-river-city-seoul-night-5830105/
- **112-113** https://pixabay.com/ko/photos/%EC%9D%BC%EB%AA%B0-%ED%95%9C%EA%B0%95-%EB%8F%84%EC%8B%9C-%EB%8C%80%ED%95%9C%EB%AF%BC%EA%B5%AD-6567167/
- **115** https://pixabay.com/photos/yeouido-han-river-seoul-bamseom-1602524/
https://pixabay.com/photos/park-recreation-outdoor-people-410256/
https://pixabay.com/ko/photos/%EC%84%9C%EC%9A%B8-%EC%97%AC%EC%9D%98%EB%8F%84-%ED%95%98%EB%8A%98-%EA%B5%AC%EB%A6%84-410260/
- **139** https://pixabay.com/zh/photos/bamboo-damyang-sunshine-364112/
- **147** https://www.istockphoto.com/kr/%EC%82%AC%EC%A7%84/%EC%95%84%EC%8B%9C%EC%95%84%EB%82%98-%ED%95%AD%EA%B3%B5-%EB%B3%B4%EC%9E%89-747-400-gm482729294-70523235
- **164** https://pixabay.com/ko/photos/%EA%B2%BD%EB%B3%B5%EA%B6%81-%EA%B6%81%EA%B6%90-%ED%95%98%EB%8A%98-%EA%B5%AC%EB%A6%84-4966220/
- **185** https://triple.guide/articles/4f0717c2-5b0a-4b8c-b8fc-028f29e512e2
- **186** https://pixabay.com/photos/nature-tree-autumn-outdoors-fall-3255573/
- **187** https://pixabay.com/ko/photos/%ED%95%9C%EA%B5%AD-%EC%A0%84%ED%86%B5-%EA%B3%A0%EA%B6%81-%EA%B1%B4%EC%B6%95-%EB%AC%B8%EC%82%B4-2777845/
- **188-189** https://pixabay.com/photos/changdeokgung-palace-palace-korea-4545145/
- **240** kakaomap(20220410)
- **241** https://pixabay.com/photos/flower-nature-white-garden-4018750/
- **253** https://pixabay.com/photos/changgyeonggung-palace-night-view-551221/
- **269** https://pixabay.com/zh/photos/traditional-hanok-palace-7540799/
- **273** https://pixabay.com/photos/lotte-world-tower-seoul-1791802/
- **279** https://pixabay.com/photos/seoul-city-landscape-architecture-7423601/

빛의 디자인
공간의 마술
ⓒ 2025. 이연소 all rights reserved

(주)유엘피
서울특별시 송파구 송파대로 167
문정동 테라타워 B동 723호 (05855)
tel 02. 322. 5932 | fax 02. 322. 5938
www.ulplighting.com

지음 이연소
기획 장희우
감수 최기수

초판 1쇄 펴냄 2025년 3월 15일
펴낸곳 픽셀하우스
등록 2006년 1월 20일 제319-2006-1호
주소 서울시 강남구 논현로 26길 42, B1 studio
제작 픽셀커뮤니케이션

*저작권법에 의하여 한국 내에서 보호를 받는 저작물이므로 어떤 형태로든 무단 전재와 무단 복제를 금합니다.

ISBN 978-89-98940-26-3 (93600)
PRICE 49,000 won

DESIGN OF LIGHTS
SPATIAL MAGIC
ⓒ 2025. LEE YEON SO all rights reserved

ULP Co., Ltd.
Urban Lighting-Design Partnership Co., Ltd.
723, Tera Tower B, 167 Songpadaero
Songpa-gu, Seoul, Korea (05855)
tel 02. 322. 5932 | fax 02. 322. 5938
www.ulplighting.com

Writer Lee Yeon So
Planning Chang Hee Woo
Advice Choi Key Soo

Published by Pixelhouse
B1 studio, 42, Nohyeonro 26 gil, Gangnamgu, Seoul
Produced by Pixelcommunication

*all rights reserved. no part of this publication may be reproduced, stored in a retrieval system, or transmitted by any means, electronic, photocopying, recording, or otherwise, without the prior permission of the copyright owner.

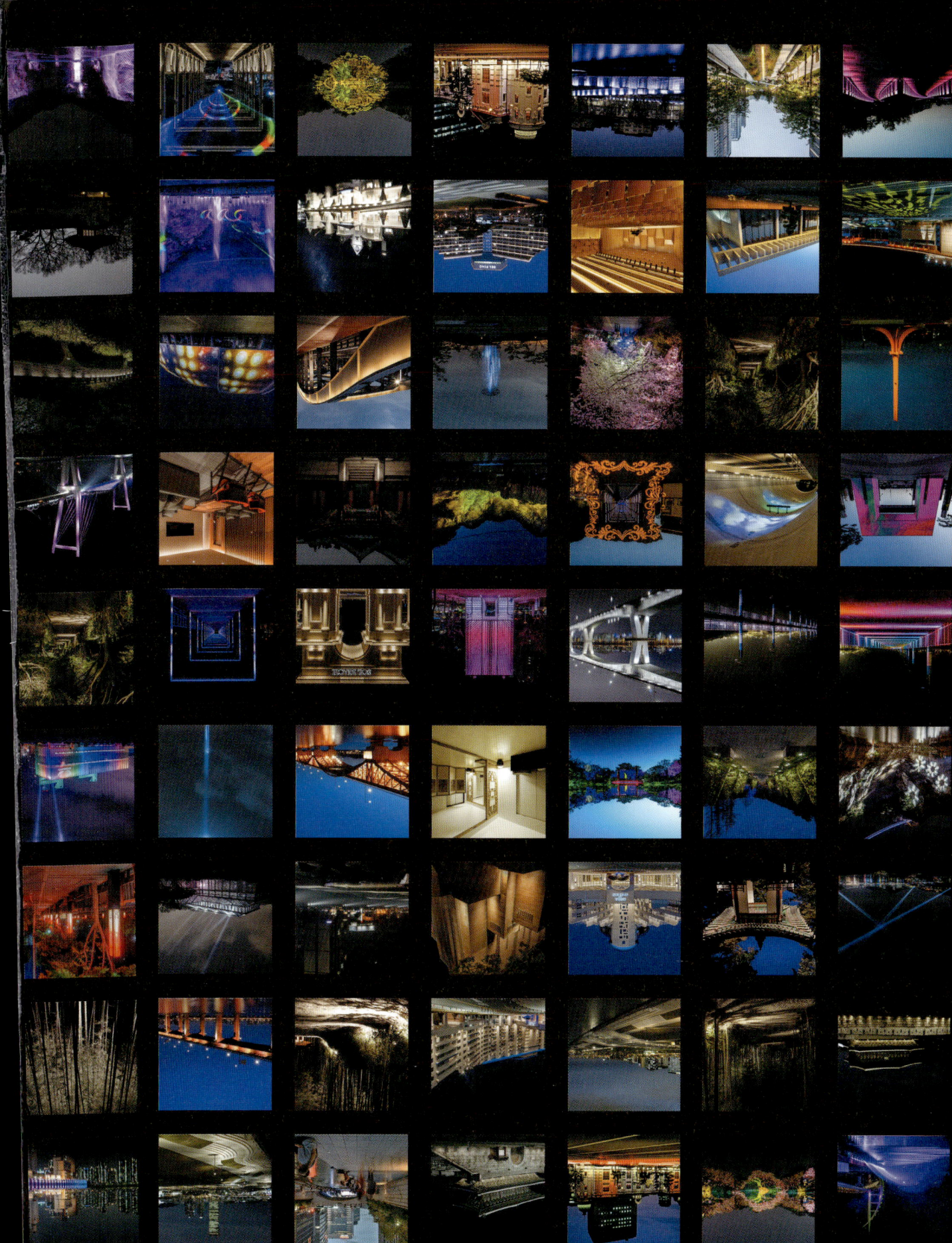